ISAAC NEWTON'S
PHILOSOPHIAE NATURALIS PRINCIPIA MATHEMATICA

ISAAC NEWTON'S
PHILOSOPHIAE NATURALIS PRINCIPIA MATHEMATICA

15-17 October 1987
Lublin, Poland

Editor

W. A. KAMINSKI

Institute of Physics
Marie Sklodowska – Curie University
20 – 031 Lublin, Poland

World Scientific
Singapore • New Jersey • Hong Kong

Published by

World Scientific Publishing Co. Pte. Ltd.
5 Toh Tuck Link, Singapore 596224
USA office: 27 Warren Street, Suite 401-402, Hackensack, NJ 07601
UK office: 57 Shelton Street, Covent Garden, London WC2H 9HE

British Library Cataloguing-in-Publication Data
A catalogue record for this book is available from the British Library.

ISSAC NEWTON'S
Proceedings of the Conference on Philosophiae Naturalis Principia Mathematica

ISBN-13 978-9971-5-0533-2
ISBN-10 9971-5-0533-9

Lublin Tercentenary Celebration

Isaac Newton's

Philosophiae Naturalis Principia Mathematica

Organized by   Department of Theoretical Physics

M. Curie-Skłodowska University

Lublin, Poland

ORGANIZING COMMITTEE

Chairman   : S. SZPIKOWSKI

Scientific Secretary: W. A. KAMIŃSKI

Isaac Newton

(1643 - 1727)

PREFACE

Three hundred years ago **Philosophiae Naturalis Principia Mathematica** directed the progress of physics toward its modern course. For years a source of inspiration not only to physicists, this seminal work become a landmark which definitively separated the new science of Latin Europe from the science of the Middle Ages still dominated by ancient authorities. Newton, whom Albert Einstein, reflecting on the bicentenary of his death, called a "bright spirit", opened up new vistas for modern thought, scientific discovery and science´s practical applications. His importance in this respect is not matched by any other thinker before and after him. Classical physics and modern onthology and cosmology have always been defined vis-a-vis the **Principia**, even when they stood in opposition to Newton´s work. But this work marks a turning point in Western intellectual history not only because of its revolutionary impact at a given moment in time. More important seems to be the fact that many thoughts formulated by Newton three centuries ago maintain their relevance even today and thus they transcend the merely historical dimension of the Cambridge scientist´s greatness.

Our celebration of the tercentenary of the publication of the **Principia** beside being a ceremonial event has yet another, perhaps less obvious but possible even more important, aspect. Quite unexpectedly to the physicists, historians and methodologists of science, philosophers and theologians gathered in the conference rooms of the Czartoryski Palace, this celebration became an occasion for the boundaries among the narrowly defined specialities of the Tower of Babel of contemporary science. It thus provided multiple opportunities for dialogue and intellectual partnership of people representing many different fields of special interest. For obvious reason, the atmosphere of agitated discussion over many unconventional ideas is reflected in the pages of this volume only in a limited sense. I hope that future meeting of physicists and philosophers in Lublin will help tc make the picture more complete by taking up new methodological and philosophical problems pervading today´s physics.

As the editor of this Proceedings and the scientific secretary of the Symposium, I want to thank all the participants and authors for their personal commitment which made the success of our meeting possible. Grateful acknowledgment should be made of the sympathetic support, also financial and material, offered by the Director of the Institute of Physics at Maria Curie-Skłodowska University and the President of the Lublin Scientific Society. I am particulary indebted to Miss T. Piasecka for her assitance and good advice in organizational matters and her work on the final version of the manuscript. I also thank World Scientific for making the proceedings of the Lublin Symposium available to readers outside Poland.

W.A.K.

CONTENTS

II.   PHILOSOPHIAE NATURALIS PRINCIPIA MATHEMATICA
Philosophical Implications

Nature and nature's laws lag hid in night
God sad: "Let Newton be!" And all was light

Alexander Pope

A Pre-Newtonian Handbook of Optics: Witelo's Perspectiva

PAWEŁ CZARTORYSKI
Institute for the History of Science,
Education and Technology,
Polish Academy of Sciences
Nowy Swiat 72
Warsaw, Poland

ABSTRACT

The paper presents Witelo's short biography, a summary of his main work, the Perspectiva, its difusion in the Middle Ages, its lasting influence till the times of Newton, and current work on its publication.

Witelo /c-a 1230/1235 - after ca. 1275/ was one of the very few representatives of exact sciences in the thirteenth century. Born in Silesia, a province of Poland, he speaks of himself as "filius Thuringorum et Polonorum", which indicates, that his father was a settler from Thuringia, and his mother a local Polish woman.

Witelo witnessed in his childhood the first Mongol invasion during which Duke Henry of Silesia from the Dynasty of the Piast Kings was killed at the battle of Legnica in 1241. Twenty years later, in 1262, Witelo went to Padua as tutor of Duke Henry's son, Włodzisław, who soon became Archbishop of Salzburg, while Witelo stayed at the University of Padua, studying canon law. In 1268, however, he joined the Papal court at Viterbo near Rome, hoping to obtain some ecclesiastical benefice. He remained there for at least three years, during a long vacancy of the Holy See, waiting for the election of the new Pope till September 1271.

Witelo's Perspectiva was most probably written during that period. Its author did not work in isolation, since Viterbo was at the time an important intellectual centre. Amongst its residents two Dominican monks should be mentionned here: Martinus Polonus, author of a Chronicle of Popes and Emperors, and William of Moerbeke, Witelo's friend, to whom he dedicated his Perspectiva. William, who knew Greek, translated into Latin some of Aristotle's works at the request of an other friend - Thomas Aquinas. William was also on good terms with the mathematician Campano de Novara and the Belgian astronomer, Henry Bates of Malines.

In 1269 William of Moerbeke translated some mathematical works of Archimedes, Eutocius, Hero of Alexandria and Ptolemy, which Witelo needed to write the first mathemathical book of the Perspectiva. This was an important early contact of Latin and Greek science. A similar cooperation between a philologist and a scientist occured in the fifteenth century between the great Greek scholar Cardinal Bessarion and the famous astronomer and mathematician Johannes Regiomontanus. It should be added, however, that the Perspectiva is essentialy based on a work on Optics by Alhazen, which at the time was already translated into Latin. Thus Witelo acquainted the Latin-speaking world with the achievements and traditions of Greek and Arabic science.

Information on Witelo's further life is scarce. Dr. Burchardt from Wrocław recently published an interesting biography, collecting available evidence, and forming new hypotheses as to the last years of Witelo's life.

In his dedicatory epistle to William of Moerbeke
Witelo presents the "methaphysics of light" in a Platonic
tradition, speaking of the order of beings, of the "lumen
divinum"in the spiritual order, and of physical light in
the material one. Howevwr, in the ten books of the <u>Perspec-</u>
<u>tiva</u> which follow, he does not refer any more to that kind
of Platonic philosophy.

Book I contains 137 geometrical propositions needed
for the understanding of the remaining nine optical books,
which are divided, in turn, into three traditional categories - books II - IV dealing with direct vision, books
V - IX with catoptrics, and book X with refraction.

To those rigorously theoretical considerations Witelo
adds the descriptions of two instruments of his own construction. He also gives a detailed anatomical description
of the eye as the organ of sight. He speaks of many simple
observations and experiments, discussing i. a. several
personal observations of the rainbow and of a halo around
the sun, as well as some optical illusions heard of in his
native Silesia. This leads him to interesting considerations concerning psychology and psychopathology of vision.

Witelo's impact on further development of the science
of optics is best presented by David Lindberg in his <u>In-</u>
<u>troduction</u> to a facsimile reprint of the <u>Perspectiva</u> /New
York-London, 1972, p. XXI, emphasis provided/:

> The optical scene in western Europe from the thir-
> teenth through the sixteenth centuries was dominated
> by the works of Alhazen and Witelo. There is scarce-
> ly a treatise on optics written after 1250 that does
> not reveal the direct influence of one or the other
> of them, and <u>no</u> <u>treatise</u> <u>at</u> <u>all</u> <u>that</u> <u>escaped</u> <u>their</u>
> <u>influence</u> <u>indirectly</u>.

4

In the fourteenth century Witelo's work was used by
Nicole Oresme, Blasius of Parma and Henry of Hesse. At
Oxford students had the choice between attending lectures
in geometry according to Euclid, Alhazen or Witelo. One
of the early lecturers at Cracow University quotes Alhazen
and Witelo, though the programme of the local Quadrivium
recommended the Perspectiva communis, an elementary handbook by John Pecham.

Extant manuscripts of the Perspectiva have been
listed by David Lindberg /op. cit./, and also by Sabetai
Unguru and A. Mark Smith in their respective editions of
books I and V of Witelo's work. Recently, Jerzy Burchardt
prepared detailed descriptions of all known manuscripts,
adding four new items to the lists of his predecessors,
namely three at the Bibliothèque Nationale in Paris, and
one at Pembrook College in Cambridge. Burchardt's descriptions still have to be published. They comprise twenty
complete and ten incomplete manuscript copies comming from
the XIIIth - XVIth centuries. Six manuscripts are preserved at the Vatican Library; two in Florence and one in
Milan; six at the Bibliothèque Nationale in Paris and one
in Dijon; two in Oxford, two in Cambridge, one in London;
finally, one in each of the following places: Berlin, Er-
furt and Munich; Basel and Bern; El Escorial and Madrid;
Vienna; Cracow.

In the sixteenth century there were two printed ver-
sions of the Perspectiva. The first one was edited by
Georgius Tanstetter and Petrus Apianus and published at
Nuremberg in 1535; subsequently it was reprinted there in
1551. The second printed version appeared in Friedrich
Risner's Opticae thesaurus published at Basel in 1572, and
reprinted in New York in 1972 with an Introduction by
David Lindberg. Risner's Preface to the Basel edition con-

tains important information on Witelo's work. It did
not lose its value till the present day.

For, according to Lindberg /op. cit./:

... the citations to Alhazen and Witelo that fill
optical texts written from about 1550 to 1650 testify to the profound impact of Alhazen and Witelo
on the foundations of modern optical theory.

Much earlier, for in the fifteenth century, Johannes
Regiomontanus owned a manuscript copy /now at Basel/ of
the Perspectiva, which he intended to publish in his
printing office; Witelo's work, at the time, was well
known to Lorenzo Ghilberti, Fra Luca Pacioli and also to
Leonardo da Vinci.

In the sixteenth century Nicholas Copernicus obtained
from his only pupil, George Joachim Rheticus, a copy of
the first edition of the Perspectiva /now at Uppsala/.
The Perspectiva was also used by Erasmus Reinhold and
Caspar Peucer, while John Dee owned a manuscript copy,
preserved at the Bodleian Library in Oxford. Witelo was
quoted by Tycho Brahe in De disciplinis mathematicis ora-
tio /1574/; the Perspectiva was also used by Michael
Maestlin, Fabricius de Aquapendente, William Gilbert and
Simon Stevin, who admitts, that his work in catoptrics
has been made "selon des écrits d'Alhazen et de Witellon".

At the beginning of the seventeenth century Johannes
Kepler's Ad Vitellionem paralipomena /1604/ and his Diop-
trice /1611/ made Vitelo's optics potentially obsolescent.
Nevertheless, interest in the Perspectiva continued. This
is true for Thomas Harriot and Willebrord Snell /the sine
-law of refraction/, for René Descartes and Galileo, who
quoted Witelo in Il Saggiatore /1623/ and in the Dialogo
sopra i due massimi sistemi del mondo /1632/, as well as

in two short papers. In Cracow, Johannes Broscius /Jan Brożek/ mentionned Witelo with great respect in Aristoteles et Euclides defensus contra Petrum Ramum et alios /1638/. Finaly, Francesco Sizi, Francesco Maria Grimaldi, and also Giovanni Riccioli in his tremendous encyclopedia of knowledge, the Almagestum novum /1651/, all mention Witelo's Perspectiva. /See S. Unguru, Witelonis Perspectivae liber primus, Wrocław, 1977, pp. 37-39./

Though quite a few contributions concerning Witelo may be found in nineteenth century litterature, serious interest in his work began with the monograph by Clemens Beaumker, Witelo, ein Philosoph und Naturforscher des XIII Jahrhunderts, Münster, 1908. Beaumker's work was followed by Aleksander Birkenmajer's Études sur Witelo, printed in Cracow, mostly in Polish, in the years 1918 - 1922, and published again in a complete French translation in "Studia Copernicana", vol. IV, Wrocław 1972. In the same year David Lindberg published his Introduction to a facsimile reprint of Risner's edition of the Perspectiva, quoted previously. This coincidence, occuring on the fourhundredth anniversary of the date of publication of Risner's edition, seems to have given impact to a revival of interest in Witelo's wor'

Thus David Lindberg contributed an important article on "Witelo" in the Dictionary of Scientific Biography, vol. 14. Eugenia Paschetto and Jerzy Burchardt published independently Witelo's letter on De causa primaria paenitentiae in hominibus et de natura daemonum, written from Padua to a friend in Silesia /"Atti dell'Accademia delle Scienze di Torino" 1975, and "Studia Copernicana", vol. XIX, 1979/. Sabetai Unguru prepared an English translation with Introduction and Commentary, and a Latin edition of the first mathematical book of Witelo's Perspec-

tiva /op. cit., "Studia Copernicana", vol. XV, 1977/.
This was followed by a similar edition of book V, by A.
Mark Smith: Witelonis Perspectivae liber quintus, "Studia Copernicana", vol. XXIII, Wrocław 1983. Two studies
by Jerzy Burchardt were published in the "Conferenze"
series /n-os 87 and 94/ of the Centre of the Polish Academy of Sciences in Rome, namely: Witelo filosofo della
natura del XIII sec. Una biografia /1984/, and La psicopatologia nei concetti di Witelo, filosofo della natura
del XIII secolo /1986/.

Finally, books II and III of the Perspectiva are
forthcomming next year in "Studia Copernicana" vol.
XXVIII. The English translation and Latin edition prepared, as book I, by Sabetai Unguru, will be accompanied
by a volume containing a Polish translation, the work of
a team of scholars at the University of Toruń: Lech Bieganowski, Andrzej Bielski, Roman S. Dygdała and Witold
Wróblewski.

NEWTON'S IDEAS AT THE JESUIT UNIVERSITIES
OF SLOVAKIA

K.A.F. FISCHER

84 Rott, 67160

F - Wissembourg

ABSTRACT

In the Habsburg-monarchy the introduction of Newton's doctrine proceeded in hand with the expansion of the heliocentric doctrine as a result of the decree of Maria-Theresia and the decrees of Pope Benedict XIV. The heliocentric orientated Jesuit physicists and astronomers presented themselves as Newtonians since the name of Newton was not included in the "Index librorum prohibitorum. The change from Cartesian ideas, which were tolerated by Aristotelian scientists, to the Newtonian doctrine did not proceed "overnight". The transitional stages can be presented through the analysis of prints of Andreas Adanyi, Andreas Jasz-linszki, Antonius Reviczki, Joannes Ivanczicz, Josephus Apponyi, Franciscus Weiss, Daniel Hersching, Josephus Kenyeres and Joannes Bapt. Horvath. "Breviarium" of the Newtonian astronomy was the first publication of this kind in the Habsburg-monarchy /Tyrnaw, 1760/.

The expansion of Newtonian as well as the heliocentric doctrines proceeded in hand in catholic culture of Central Europe as a result of two decrees: the decree of Maria-Theresia of July 25, 1752, which abolished the Aristotelian doctrine at the universities of the Habsburg-monarchy, and the decree of the Pope Benedict XIV. of April 16, 1752 concerning the abolition of the geocentric doctrine[1]. But the latter date has not been clarified explicitly up to the present day. Many scientists read the year "1757".

The records, however, on page 128 had been before 1750. I want to leave this question open. I quote the words from the papal decree: ... "omitatur Decretum, quo prohibentur libri omnes docentes immobilitatem solis et mobilitatem Terrae"... Nevertheless we can see, that the lessons in mathematics and physics were reorganised at the Jesuit universities in 1749 — that means prior to these decrees. The heliocentric doctrine was emphasised, although it was admitted to be only a possible hypothesis until the suppression of the order in 1773. The Piarists and Benedictines, on the contrary, had included the heliocentric doctrine into the lectures of their schools of higher education before 1760. This process did not happen suddenly — transitional phases can be observed.

"Breviarium" of Newton's astronomical doctrine was published in 1760 at Tyrnav university[2] as the first publication of this kind in the Habsburg-monarchy. But HAHLHEIMER of the Austrian National Library in Vienna found out[3] that this was an anonymous pirate-edition of the work by Abbé Pierre SIGORNE from Paris, 1749[4].

The change in Cartesian thought, which was acknowledged by Aristotelians, toward Newton's doctrine did not take place suddenly in Slovakia. I demonstrate this process by the following examples of works by some Jesuits teaching physics and astronomy at Tyrnaw university:

Andreas ADANYI, SJ[5] was familiar with Newton's doctrine, but he rejected it taking into consideration the problem of the "exploration of causality" and he did not accept Newton's gravitational theory. To his successor Andreas JASZLINSZKI, SJ[6] Newton's gravitational theory seems to be the same hypothesis as the Peripathical-occultism was. He quotes MUSSCHENBROEK[7] as well as s'GRAVESANDE[8] almost word by word but in the fundamental evalution we come to different oppinions:

a/ in my opinion he did not solve the question, but he
   left it open.
b/ according to CSARODI[9] Jaszlinszki was an anti-Newton-
   ian.

Newton's doctrine is more deeply rooted in the works
byAntonius REVICZKY, SJ[10]. For him the "pressure of aether"
is the reason for gravitation of bodies. And he as well
as Descartes believed that an "empty space" is not possible.
But in his work "Elementa philosophiae" he explained Newton's doctrine thoroughly. Joannes IVANCZICZ, SJ[11] had
similar opinion which might be even more conclusive.
The lessons by Josephus APPONYI, SJ[12] can be described
as "Newtonian".

The facts mentioned before show that in Slovakia at
the university of Tyrnaw as well as the university of
Kashaw[13], which can be considered as a branch of Tyrnaw –
there were pros and cons concerning the Newtonian revolution. Nevertheless, we do observe, that the Jesuits in the
whole Society of the 18th century, who were oriented toward heliocentrism, presented themselves as Newtonians, because the "Oboedientia" did not allow them to quote Copernicus, who was on the "Index librorum prohibitorum" dating
from the 16th and 17th centuries. Newton's name was not
included there. Was this an alibi only?

The position of Carolus SCHERFER, SJ[14] was dominant
in the thought of the Central European provinces of the
Society. Scherfer accepted Newton's doctrine unconditionally, when he was a professor "Physicae generalis et
particularis" in Vienna 1752/53. About 1760, when the
Tyrnaw edition was printed, Scherfer was a professor of
"applied" Mathesis – which means of Astronomy in Vienna
and he was a close collaborator to Franciscus WEISS, SJ[15].
Weiss, being the director of the observatory of Tyrnaw,
did not enter these controversies. But he was a pure

experimentalist in the Newtonian sense of the word.

The "oscillations-theory of aether" was opposed for the first time in 1763 by Josephus KENYERS[16], although that heliocentrism was a possible hypothesis. He also believed that power is proportional to the square of distance. Daniel HERSCHING[17] spoke out more clearly - he did not want to carify the hypothesis in a philosophical way, but he studied natural laws.

Although Joannes Bapt. HORVÁTH, SJ[18] left us several editions of his "Physica", every edition is a completely revised book. HELL, BOSCOVICH, WEISS, SAINOVICH and other Ex-Jesuits after the suppression of the Society were practically "over-night" heliocentric, instead of being geocentric orientated. In Horváth's work we observe an "internal struggle" against the "Oboedientia" and the transition to clear heliocentrism. But Horváth was a Newtonian at the beginning - he opposed the Cartesian method "à priori" and stood for Newton's method "à posteriori".

REFERENCES

1) Archives of "Propaganda Fidei", Roma: Acta sacrae Indicis Congregationis ab Anno 1749 ad Annum 1763: here pag. 129; Die 16.Apr.1752 /1757?/, § 2., Quod, habito verbo cum SS^{mo}D.N. omitatur decretum quo prohibentur libri omnes docentes immobilitatem solis et mobilitatem terrae...

2) Astronomiae Physicae juxta Newtoni Principia BREVIARIUM, methodo scholastico ad usum Studioae Juventutis. Tyrnaviae, Typis Academicis Societatis Jesu, Anno MDCCLX.

3) Mrs. Antonia HAHLHEIMER, personal information.

4) Pierre SIGORNE, Abbé, Dr.of Sorbonne: 25.X.1719 Rambercourt - les Pots / Lorraine-10.XI.1809 Lyon.

12

Institutiones Newtoniennes. I./II. Parisiis 1748, II.Ed.
1769.
Astronomiae physicae juxta Newtonis principia breviarium.
Parisiis 1749.

[5] Andreas ADANYI, SJ: 28.XII.1716 Darmaud/Comit. Heves -
13.X.1796 Gran.
Philosophia naturalis, pars I., seu physica generalis.
Tyrnaviae 1755; pars recentior physicas, ibid. 1756

[6] Andreas JASZLINSZKI, SJ: 1.IX.1715 "Szinensis"/Hung.,
- after 1772.
Institutiones Physicae generalis et particularis. Tomi
II. Tyrnaviae 1757.

[7] Peter van MUSSCHENBROEK: 14.III.1692 Leiden - 19.IX.
1761 ibidam.
Elementa physicaes. Lugduni Batavorum 1729; ibid. 1734.
in German: ibid.1729; in French: ibid. 1739.

[8] Wilhelm Jacobus s'GRAVESANDE: 26.IX.1668 s'Hertogen-
bosch - 28.II.1742 Leiden.
Physices elementa mathematica, experimentis confirmata
ad philosophiam Newtonianam. Tomi II. Lugduni Batav.
1720-21. Supplement 1725.

[9] CSAPODI, Csaba: Newtonianismus a nagyszombati jeszuita
egyetemen. "Regnum" /Budapest/ 6.,/1944/46/, 59-68,
here pag.63.

[10] Antonis REWICZKY, SJ: 17.I.1723 Ujély/Hung. XII.1781
Preβburg.
Elementa philosophiae naturalis. Pars I., seu Physica
generalis ad usum auditorum conscripta. Tyrnaviae 1758.
Institutiones physicae Pars II., ibid.1758.
Elementa philosophiae rationalis seu Institutiones logi-
cae, metaphysicae et Theologiae naturalis. Tomi II.
Tyrnaviae 1757-58.

Although Rewiczky let "open" the Carthesian doctrine in his two works, he turns to it in his third work.

[11]Joannes IVANCZICZ, SJ: born: 25.XI.1722 Kommern - after 1772.

    a./ Universae Matheseos brevis Institutio. Partes III. Tyrnaviae 1752-1755.

    b./ Assertiones ex universa Philosophia...inaalma...Universiate Tyrnaviensi... 1759... s.l.n.d.

Also Ivanczicz turn to the Cartesian doctrine in his last work too.

[12]Josephus APPONYI, SJ: 10.X.1718 "Ungarus"-14.X.1757 in Silesia.

    a./ Apponyi praeside - Joannes MITTERPACHER de MITTENBERG propugnat: Dissertatio physica de corpore generatim, deque opposito eidem in vacuo. Tyrnaviae 1753.

    b./ /KÉRI, Franciscus, Borghia, SJ/: Dissertationes physicae... dum assertationes ex universa philosophia in ... universitate Tyrnaviensi ... 1754 ... publice propugnaret Josephus MAJLÁTH ... ex praelectionibus Josephi APPONYI, Francisci WEISS, Adami WITTMANN. Tyrnavise 1754.

    N.B.: Adamus WITTMANN could not be found out as a Jesuitprofessor of Tyrnaw in ARJI, Rome.

[13]The university of Kaschaw can be considered as a branch only of the Tyrnaw university. Meanwhile the teaching and research was pursued in Tyrnaw, the practice and education of clergy and public servants was pursued in Kaschaw. The level was the same of the both universities. We state the delay of 2-4 years in teaching the new opinions. The manuals of Tyrnaw were used in Kaschaw only little tectbook were written there.

[14]Carolus SCHERFER. SJ: 3.XI.1716 Gmunden-25.VII.1783 Vienna.

14

Institutionum physicae partes II. Viennae 1752. 3.Ed.
1768.

15) Franciscus WEISS, SJ: 16.III.1717 Tyrnaw-1780 Buda.
1752-1777 Prof.Matheseos et Astronomiae and director
of the observatory of Tyrnaw. Probably the initiator
of the publication of the "breviarium" of Pierre Sigorne.

16) Daniel HERSCHING, SJ: 1732 Fünfkirchen - after 1772.
Assertationes ex universa philosophia, in quas in ...
univ. Tyrnaviensi ...A. 1769 ... suscepit Ladislaus
RUDNYÁNSZKY ... ex praelectionibus Danielis HERSCHING
... s.l.n.d.

17) Josephus KENYERES, SJ: 24.V.1724 Steinmanger/Hung. -
after 1772 as canon of Rosenau.
Materia Tentaminis publici ex praelectionibus Physicis
Jos. Kenyeres... quod corm admodum P.Nicolai BENKÖ ...
Tyrnaviae 1763.

18) Joannes Bapt. HORVÁTH, SJ: 13.VII.1732 Gran - 20.X.
1799 Buda.
Physica generalis. - Physica particularis.
Ist Ed. Tyrnaviae 1770, last posthumous 1817 Budae.
Tentamen publicum e praelectionibus physicis. Tyrna-
viae 1771.

THE NEWTONIAN CONCEPT OF SPACE AND TIME

Stanisław Mazierski

Catholic University of Lublin

Home address: 7, Nowotki, 20-039 Lublin, phone: 381-16,
POLAND

ABSTRACT

Ancient Greeks did not create a clear theory of space
and did not distinguish the concept of place from the
concept of space. They had different notions of time.
In the trend of realistic philosophy (Aristotle) time
was closely connected with motion, in the idealistic
trend (Plato) it was considered to be the image of
eternity. I. Newton fluctuated between empiricism and
metaphysics. As a physicist he tried to infer his statements from phenomena, but in practice he went beyond
the limits of experience. He accepted the concept of
absolute time and absolute space. To prove the existence of absolute space he performed the experiment with
a rotating bucket filled with water. A. Einstein rejected the concept of absolute space and absolute time
as he considered it to be inconsistent with the Newtonian system of physics.

The paper will begin with some historical remarks presenting the main ideas concerning space and time, especially
those of ancient Greeks.

In ancient philosophy an unmistakable described theory
of space can be found. The Greek idea of space is not clear.
The Greeks did not set the notion of place against the notion
of space. If we say that Democritus understood space as an
absolute vacuum, this statement is, in fact, a postulate
resulting from his atomistic theory. It says that bodies

consist of indivisible atoms moving about in a vacuum. Because the number of atoms in the world is unlimited, their
container (a vacuum) must also be boundless.

It was Aristotle who analysed the notion of place in
his Physics and even gave it a definition: a place is delimited by the surface of the body surrounding it[1]. As the
reference system used in modern Physics was unknown to him,
he referred to the bodies which delimited places by their
surfaces. Although in his analysis the notions of place and
space are nct opposed as two different values, in his speculations ccncerning the cosmos the famous Stagirite implicitly suggests the existence of anisotropic space. In this
real space bodies reach their natural places. This phenomenon made Aristotle accept the idea of a finite world, for
in an infinite vacuum there would be no basis for thinking
that there are natural locations, that the motion of bodies
has a direction and an end.

Aristotle gave more consideration to the problem of
time. The basis and the starting point of his theory of time
is local motion, also called mechanical[2]. The understanding
of motion is very broad. Every change (local, quantitative,
qualitative) is motion and is defined as a transition from
a potency to an act. Local motion is a continuous realization of potentiality. The analysis of a spatial quantity
(a distance) and motion resulted in the following definition
of time: "Tempus est numerus motus secundum prius et poste-
rius"[3] which, when simplified, may sound as follows: time
it counting successive changes.

According to Aristotle time is eternal. To justify this
statement Aristotle referred to the circular motion in the
firmament as to the most perfect of all motions[4]. If this
circular motion exists, and it is continuous, undisturbed,
uninterruptec and unrestricted, infinite time connected
with this motion must also exist. Aristotle's historic con-

tribution to philosophy was the conclusion that time and motion are inseparable: if there is no motion, there is no time.

The above definition of time aroused much controversy and criticism in later centuries, also in the present age. Admittedly, the idea that time and motion (changes) are inseparable has, so far, resisted the criticism of the greatest thinkers. Thanks to it Aristotle is included among these philosophers who reject the concept of absolute time.

The Stoics' position was different. They referred to the concept of vacuum put forward by Democritus maintaining that the cosmos is surrounded by it from all sides. If this proposition were true, why does the whole cosmos not disperse and disappear in time? The Stoics gave an interesting answer to that question. It is so because individual parts of the material world are connected with one another by means of forces originating in the pneuma. This connecting force keeps the world together. The vacuum, devoid of all forces, cannot break these bonds. There are no inert bodies, all are active, and their activity is due not the forced mechanical motion but to the pneuma. It is the pneuma which penetrates the matter, like fire penetrates incandescent iron. The origin of motion and life is in the matter itself and not outside it. The stoic conception of the world was dynamistic as opposed to the mechanistic conception of Democritus.

Another important trend in ancient philosophy was idealism initiated by Plato. Plato maintains that time is an image of eternity and that eternity is an unchanging reality. Time is not eternal because it was created by the demiurge. It was this divine creator of the world who called time into being together with the stars (the firmament) and the planets to make the cosmos more perfect. The heavenly bodies are necessary for time to exist and the regularity of their movements is the basis for the measurement of time. If the bodies did not exist and change, we would not be able to experience time and express it in numbers. Plato's theory of time was influenced by the Pythagorean cosmological conceptions. For the author of Timaios nature was an imitation of eternity and thus, he considered that time made the world similar to eternity.

The Platonic conception of time was extended, completed and developed by the Neo-Platonists. It was Plotinus (204--269), one of the outstanding advocates of Plato's ideas, the author of Enneads who separated time from eternity. The attribute of eternity belongs to the Absolute which is more perfect than an idea and, as eternal, independent of anything. What is more, it is the origin of everything that exist. It is outside the scope of what can be comprehended with the mind. As a perfect unity (unitas perfecta) and infinity it goes beyond the frontiers of cognizability. The separation of eternity and time was the source of the idea that eternity, in a strict sense, is an attribute of God who is the most perfect being, and that temporariness is an attribute of all transitory being. From this point of view the Absolute is a timeless being (Boethius. Thomas Aquinas).

The idea of absolute space and absolute time came into view also in modern times. It found its followers in the persons of the French philosopher and mathematician P. Gassendi (1592-1655) and the author of classical physics I. Newton (1642-1727).

The concept of space developed in the 17th century may be characterized as follows: absolute space is a boundless container for material substances; it is a result of abstraction, an image of a three-dimensional extension without limits. It was called absolute because it was believed that it is independent of time, of the bodies moving about in it and, at the same time, necessary for the existence

of material objects which were, are and will be. The philosophers who maintained that such a space could exist all by
itself, i.e. without any bodies ascribed various properties
to it. Absolute space is (a) an infinite and boundless container of bodies, (b) it never contracts or expands, for a
contracted or expanded space would be inconsistent with the
attribute of infinity, (c) it is homogeneous and not differentiated, (d) it is motionless as a whole and in its parts,
for if it were capable of moving, it would move about in
something else, and this is impossible, for there is no
other reference system except absolute space, (e) it is in-
destructible, not created and eternal; we cannot imagine
that it does not exist. We can imagine space without bodies,
but we cannot imagine bodies without space. The assumption
that material substances can be annihilated does not contain a contradiction, whereas the hypothesis that space can
be annihilated is absurd.

Does such an imaginary space have its equivalent in
reality?[5] It was P. Gassendi, a continuator of the ideas of
mechanicism professed by Democritus and Epicurus, who believed that absolute space exists and would exist even if
bodies were annihilated. Although it is neither a substance
nor a property, it is something really existing, different
from and independent of bodies and all changes. The universe is spatially infinite[6].

If absolute space is neither a substance nor a property, then the question arises - what is it? There were
philosophers before Gassendi who maintained that space was
a container for all bodies and changes taking place in it.
Gassendi went even further in his speculations distinguishing (1) the three-dimensional extension of bodies from (2)
the three-dimensional extension of space. The former is material, the latter immaterial in the sense that it is not
constituted by material objects and can exist without them.

The concepts of time and space were undergoing   many
changes from the times of ancient Greeks to   modern times,
but these changes were not radical. Time and space were
objects of philosophical speculations. Newton   introduces
these concepts into his work Philosophiae Naturalis Principia Mathematica, but not uncritically. He reflects   upon
them within the scope of his theory of cognition[7]. On the
basis of his statements included not only in Principia Mathematica we can distinguish two, as it were, spheres   of
cognition. One of them is a set of material objects, phenomena and processes which can be examined experimentally.
This examination includes observation, inferring from phenomena and induction[8]. The other is a sphere which cannot be
contained within the framework of empirical studies   but
which is brought to mind in the wide context of physical studies. The objects of the other sphere are the   results of
man's intellectual speculation which includes metaphysical
theses, hypotheses not inferred from phenomena, theses of
religious character, but not revelational, statements concerning the harmony and the beauty of nature. The author of
classical physics did not only create mechanics, but   also
referred to the views of those metaphysicists who tended towards Platonism. Intellectual affiliations with Greek philosophers resulted in Newton's heterogeneous position   in his
theory of empirical cognition. Although he maintained   that
it was a fallacy to believe that every phenomena in nature
serves a purpose, because phenomena and processes   in   the
world are governed by the laws of mechanics, it did not prevent him from voicing the argument for the existence of God,
known under the name of "the physical-teleological argument".
    Newton was a physicist and a thinker fluctuating   be-
tween empiricism and metaphysics[9]. Evidence for this can be
found in his writings and letters to his friends.   In   the
letter to R. Bentley (1693) in which he makes several re-

marks about gravity as an important and internal force of
the matter, Newton asks not to be credited with this idea.
The knowledge of the causes of gravity is not a discipline
for a physicist. He maintains, however, that gravity must
be caused by some factor, but the question whether it is of
material or immaterial nature is left unsolved.

Trying to interpret the above mentioned remarks we
could express them as follows: I can tell you with what
force (F) two bodies (m_1, m_2) situated at a distance (r)
from each other are mutually attracted. This force can be
expressed by the formula $F = \frac{m_1 m_2}{r^2}$, but do not ask me about
the cause of this force of gravity. The position of the
author of classical physics is understandable. He points to
this field of physical studies which can be explored by
means of empirical methods. What goes beyond this sphere of
cognition belongs to the realm of metaphysics.

Towards the end of Scholium Generale, reflecting upon
his considerations, Newton writes: Hactenus phaenomena coe-
lorum et maris nostri per vim gravitatis exposui, sed causam
gravitatis nondum assignavi. Oritur utique haec vis a causa
aliqua, quae penetrat ad usque contra Solis et planetarum,
sine virtutis diminutione. Later in this context he remarks:
Rationem vero harum gravitatis proprietatum ex phaenomenis
mondum potui deducere, et hypotheses non fingo. Quidquid
enim ex phaenomenis non deducitur, hypothesis vocanda est
... Et satis est quod gravitas reveri existat, et agat secundum leges a nobis expositas, et ad corporum coelestium
et maris nostri motus omnes sufficiat[10].

In Scholium Generale Newton states that he explained
heavenly phenomena, i.e. the motions of planets and comets,
high and low waters by means of the force of gravity. This
force exists, but has other properties than mechanical forces. He stated however, that the cause (ratio) of the force
of gravity could not be inferred from phenomena and that he

formed no hypotheses.

In this way the two above mentioned spheres of cognition were separated by a demarcation line in the form of the statement "hypotheses non fingo".

Newton did not always observed this demarcation line and went beyond the frontiers of the above mentioned spheres of cognition. Evidence for this is the fact that he accepted the concepts of absolute space and absolute time. At the beginning of Philosophiae Naturalis Principia Mathematica (in Scholium Generale) Newton remarks that the concepts of space, time and motion are known to everybody from everyday experience and that is why they need not be explained, still he analyses and explains them.

Absolute space is, by nature, not related to anything external, is always the same and motionless[11]. Relative space is a measure of absolute space or an extension (dimensio) determined by the positions of bodies. Similarly, the author of classical mechanics distinguishes between an absolute place and a relative place. A place is part of space occupied by a body and according to space it is either absolute or relative[12]. There also exist absolute motion and relative motion. We can speak about absolute motion when a body moves from one absolute place to another absolute place, when a body moves from one relative place to another relative place the motion is relative[13]. It follows that absolute motion is related to absolute space and relative motion to other bodies, i.e. to relative space. Similarly, absolute time, genuine, mathematical, by its very nature and in itself flows evenly and uniformly, irrespective of any external object[14].

Newton's absolute space and absolute time are not only products of philosophical speculation as was the case with his predecessors. They are the basis of his philosophy of nature and the necessary condition of his philosophical sys-

tem in which the laws of motion - together with force, mass
and gravity - play an important part. Newton considered the
laws of motion to be unquestionable experimental principles.
Today we know that, e.g. the first law of motion is not em-
pirically verifiable: If no force acts upon a body, this
body is at rest or moves rectilinearly and uniformly. To be
more precise, there is no body which is not affected by a
force and which is at rest. We cannot create such physical
conditions in which the first laws of dynamics could be
verified experimentally. All reference systems are also in
motion. Newton treats all laws of motion as "axioms" (axio-
mata sive leges motus). They serve as the basic assumptions
of physics from which theses are inferred which are later
verified experimentally. Formulating these laws Newton, in
a way, goes beyond the frontiers of concrete experience,
but at the same time shows the direction of the development
of natural sciences. If we confined ourselves to the limits
of experience, we would impoverish empirical sciences.

Although the famous formula"hypotheses non fingo" was
used for the explanation of gravity, it later became a motto,
a methodological principle ordering to exclude such notions
as concealed properties and other non-empirical notions.
Newton did not negate metaphysical theses, did not consider
them to be absurd. He only separated them from empirical
statements. He did not reject the existence of real objects
which go beyond the limits of human experience. He merely
maintained that they had no connection with natural studies,
with one exception - absolute space. As a believing Christian he considered it to be a"Sensorium Dei"which helped
him understand God's omnipresence in the universe. That was
a definite turn towards metaphysics.

Newton accepted the concept of absolute space, but unlike his predecessors, he did not limit himself to philosophical speculations in that matter. As he wanted to be faith-

ful to his methodological principle "hypotheses non fingo",
he tried to prove the existence of absolute space experimentally. He knew that absolute space was not within the
range of our senses so, he tried to prove its reality by
inference from the properties of absolute motion. While Greek
philosophers declared that space was absolute and that absolute motion was moving from one absolute place to another,
giving no arguments in support of their position. Newton
adopted a different procedure. First, he looked for the signs
of absolute motion and only from this phenomenon he inferred
about the reality of absolute space. The fact that Newton
adopted this procedure is another proof that the author of
classical physics fluctuated between the two spheres of cognition: empirical and metaphysical.

The existence of absolute space was suggested to Newton by the experiment with a revolving bucket full of water
which he concucted himself[15]. The analysis of this experiment led him to the conclusion that there existed a connection between absolute motion and absolute space. Showing
that in nature there are absolute motions taking place in
relation to motionless, absolute space would be a proof for
the existence of this space. Because a uniform motion of a
body in relation to absolute space cannot be detected, absolute motion must be variable (with acceleration). Bodies
moving about in inert systems are affected, besides the giv-
en forces, by the so-called forces of inertia.

In the above mentioned experiment the molecules of
water rotating together with the bucket are affected, besides the force of gravity, by the centrifugal force of
inertia. As the result of the vectorial sum of these forces
the surface of the water assumes the shape of a revolving
paraboloid. If it were not for the centrifugal force, the
surface of water in the rotating bucket would be flat. The
fact that a moving body is affected by the force of inertia

leads to the conclusion that the motion of this body is absolute in relation to absolute space.

According to the principle of the relativity of motion it does not matter whether the bucket revolves in relation to the earth, the stars (the universe), or whether the universe revolves in relation to the bucket. In both cases the physical effect should be the same. According to Newton the forces of inertia violate this principle. When the universe the surface of the water in the bucket is raised by its walls. Therefore, the rotary motion of the bucket is absolute.

Following Newton's line of reasoning we would have to admit that the thesis "system A moves with acceleration in relation to system B" and the thesis "system B moves with acceleration in relation to system A" are not equivalent because they describe different physical situations. System B, in which there act no forces of inertia, is at absolute rest in relation to motionless absolute space. In the accelerated system the forces of inertia are of great importance and are responsible for the process taking place in this system.

It is well-known that Ernest Mach criticised the concept of absolute motion of accelerated systems. He tried to replace the Newtonian concept of absolute motion of accelerated systems responsible for absolute acceleration with the principle: "everything in nature can be explained by the interaction of masses" Also Einstein rejected the thesis that absolute empty space affects the behaviour of bodies because it is incompatible with the essential content of the Newtonian system of physics.

N o t e s

[1] Phys. IV, 4, 212 a 20.

[2] Phys. IV, 11, 218 b 12-33.

[3] Phys. IV, 11-14, 219 b 1-2. Defining time as the measure
of changes (numerus motus), or counting successive changes Aristotle distinguished the objective aspect of time
- motion and the subjective aspect - the mind counting
these changes. Without counting changes we are left with
only the substrate of time, i.e. motion.

[4] Phys. VIII, 8, 265 a 8-12.

[5] Fr. Suarez (1548-1617), Disputationes metaphysicae, XXX,
p. 7, n. 17; LI, p. 3, n. 12, n. 23.

[6] P. Gassendi (1592-1655), Exercitationes paradoxicae ad-
versus Aristotelem, 1624; Syntagma philosophicum complec-
tens logicam, physicam et ethicam, 1658.

[7] I. Newton, Philosophiae Naturalis Principia Mathematica,
Vol. I-II, Glasguae 1833.

[8] Omnis enim philosophiae difficultas in eo versari vide-
tur, ut a phaenomenis motuum investigemus vires naturae,
deinde ab his viribus demonstremus phaenomena reliqua.
Ibidem: Praefatio ad Lectorem, XI.

[9] Heimo Dolch, Kausalität im Verständnis des Theologen und
der Begründer Neuzeitlicher Physik, Verlag Herder Frei-
burg 1954, pp. 139-167.

[10] Principia Mathematica, Vol. II. pp. 201-202.

[11] Spatium absolutum, natura sua sine relatione ad externum
quodvis, semper manet similare et immobile: relativum
est spatii huius mensura seu dimentio quaelibet mobilis,
quae a sensibus nostris per situm suum ad corpora defini-
tur, et a vulgo pro spatio immobili usurpatur: uti dimensio spatii subterranei, aërei vel coelestis definita per
situm suum ad Terram. Ibidem, Vol. I, pp. 8-9.

[12] Locus est pars spatii quam corpus occupat, estque pro ratione spatii vel absolutus vel relativus. Ibidem.

[13] Motus absolutus est translatio corporis de loco absoluto in locum absolutum, relativus de relativo in relativum. Ibidem, Vol. I. p. 9.

[14] Tempus absolutum, verum, et mathematicum, in se et natura sua sine relatione ad externum quodvis, aequabiliter fluit, alioque nomine dicitur duratio: relativum, apparens, et vulgare est sensibilis et externa quaevis durationis per motum mensura ... Vol. I, p. 8.

[15] This description of the experiment with the rotating bucket, filled with water, is also included in the Scholion, Vol. I, pp. 12-14. Vide also: C. Kittel, W. D. Knight, M. A. Ruderman, Mechanics, New York 1965, Chapter 3.

THE ROTATION IN NEWTON'S WORDING OF HIS FIRST LAW OF MOTION

Martin Černohorský

J. E. Purkyně University of Brno
Faculty of Natural Sciences
Department of General Physics

Kotlářská 2, 61137 Brno, Czechoslovakia

ABSTRACT

The article advances the opinion that Newton's Principia and Newton's manuscripts furnish evidence that the Latin wording of Law I in the Principia expresses the principle of inertia for both the translational and rotary motions. Five items of evidence are given successively in sections 4-8 of the following contents.

1. Characterization of the problem
2. Ambiguities in Newton's First Law of Motion
3. Newton's wording of Law I in the Principia
4. Newton's translation/rotation commentary on Law I
5. Newton's explicit embodiment of the rotation in Law 1
6. Newton's change-over from univocal "in linea recta" to equivocal "in directum"
7. Newton's change-over from six laws to three axioms
8. Newton's reference to Law I in the case of pure rotation
9. Three phases of Newton's formulations of the principle of inertia
10. A survey of evidence for the translation/rotation interpretation of Law I

1. CHARACTERIZATION OF THE PROBLEM

The first of Newton´s Axioms, or Laws of Motion often is a subject of cogitations on its logical justification (whether it is not superfluous being easily deducible from the Second Law, or following from the preceding Definitions, e. g. /5/, p. 240; /6/, p. 325), or about its physical meaning (e. g., whether it should not be taken just as a definition of the inertial reference frame, e. g. /7/). In doing so, there appeared from the first translation /2/ of the Principia through all the time up to the present no doubt about the veracity of the linguistic interpretation of the original Latin wording.

However, it seems certain that the translations in use in the physical literature, as far as they are declared as interpreting Newton´s own wording, omit several facts, and are in want of a revision. To demonstrate this assertion two arguments are offered in evidence in the Principia itself, and another weighty support is to be found in Newton´s manuscripts /4/.

Meditating on the matter, it must be taken into account that Newton did not look on a uniformly rotating body as a special case of the law of conservation of the angular momentum which might be deduced from the Second Law, in accordance with the present practice, but as known to be true from experience. Furthermore, dealing with the interpretation of the original Latin wording of Law I, the following two problems should be strictly distinguished: (i) The interpretation of Newton´s 1687 wording of the principle of inertia. (ii) The wording of the 1987 interpretation of the principle of inertia.

The problem (i) will be dealt with, whereas the problem (ii) is not considered here. The above interdistinction is essential. There is no aim to treat the problem

(i) in the connection with the problem (ii) in any place of the following text. The sole intention is to indicate evidence that Newton worded Law I while having in mind to formulate the principle of inertia for both the translational and rotary motions.

Note: Wherever Law I or Lex I appears, the first law of motion is meant in the formulation to be found in the Principia (/1/, p. 12; /2/, pp. 13, 644; here section 3); and wherever Law 1 or Lex 1 appears, the first law of motion is meant in the formulation to be found in a manuscript preceding the Principia (/4/, pp. 307, 312; here section 5).

2. AMBIGUITIES IN NEWTON´S FIRST LAW OF MOTION

Newton´s First Law of Motion entered into the physical literature in the sense corresponding to the following wording (/2/, p. 13):

"Law I. Every body continues in its state of rest, or of uniform motion in a right line, unless it is compelled to change that state by forces impressed upon it."

Thus, Newton´s Law I is universally understood as enunciated only for the translational motion.

However, such an interpretation of Newton´s original Latin wording is hardly compatible with his commentary on Law I presenting examples of pure translation, pure rotation, and superposition of translation and rotation /8/. Another ambiguity is that such an interpretation makes Law I only a special case of Law II and the axiomatization seems to be not accomplished (e. g. /5/, p. 240). Furthermore, with the wording as being meant only for the translational motion, Newton´s reference to Law I in the case of pure rotation is incomprehensible /10, 11/.

All of these ambiguities disappear by making use of the equivocal meaning of the Latin expression "in directum". The successive sequence of Newton's more and more elaborate wordings of the principle of inertial motion (/4/, p. 29-30) furnish evidence that at last Newton formulated his Law I in the Principia having in mind both translational and rotary motions. His Latin wording with the equivocal "in directum" makes possible an interpretation corresponding to this translation/rotation conception.

3. NEWTON'S WORDING OF LAW I IN THE PRINCIPIA

There are two slightly different wordings of Law I in the Principia (/2/, p. 644).

Editions 1687 and 1713:

"Lex I. Corpus omne perseverare in statu suo quiescendi vel movendi uniformiter in directum, nisi quatenus a viribus impressis cogitur statum illum mutare."

Edition 1726:

"Lex I. Corpus omne perseverare in statu suo quiescendi vel movendi uniformiter in directum, nisi quatenus illud a viribus impressis cogitur statum suum mutare."

The translations by Motte (/2/, p. 13) and by Thomson and Tait (/3/, p. 241) correspond rather to the First and Second Editions (statum illum mutare - to change that state), although according to /2/, p. 638, and to /3/, p. 241, they were made from the Third Edition (statum suum mutare - to change its state); this fact is, however, irrelevant to the translation/rotation problem.

Translation by Motte /2/

"Law I. Every body continues in its state of rest, or of uniform motion in a right line, unless it is compelled to change that state by forces impressed upon it."

Translation by Thomson and Tait /3/

"Law I. Every body continues in its state of rest or of uniform motion in a straight line, except in so far as it may be compelled by force to change that state."

However, the expression "in directum" does not necessarily mean only "in a right line" or "in a straight line". The translational motion with a constant velocity is a uniform motion taking place in a given direction; and the rotary motion with a constant angular velocity, too, is a uniform motion taking place in a given direction (e. g. the Earth rotates "forwards" in the given direction from W. to E.). Thus, a uniform motion in the given direction means either a uniform translational motion in a straight line (the velocity is a constant vector), or a uniform rotary motion about an invariable axis (the angular velocity is a constant vector), or the superposition of both.

The vectorial aspect stressed above in the brackets is in accordance with Newton´s dynamical thought presented in /4/, Chapter 5, section 5.3. Dynamics of a Single Rotating Body, p. 82:

"Combination of circular motions

Imagine a body rotating with angular velocity R about an axis AC acted on by a new force which if applied alone would cause the body to rotate about another axis CB with angular velocity S . Newton then gives a rule for finding the new axis of rotation of the body and its angular velocity about it. Although the rule only holds if the moments of inertia of the body about the three axes in question are equal, it is interesting for the indication it affords of Newton´s realization of the vector nature of angular momentum."

Having all the facts mentioned above in mind, the translation of Newton´s wording of Law I may read as follows:

<u>Law I. Every body continues in its state of rest or</u>
<u>of uniform motion in the given direction, unless it is</u>
<u>compelled to change its state by forces impressed upon it.</u>

The interpretation of this wording for the translational motion is apparent, whereas for the rotary motion
it is not only strange, but also rather strained. However,
it is consistent with the otherwise hardly explainable
facts indicated in the next sections 4-10.

4. NEWTON´S TRANSLATION/ROTATION COMMENTARY ON LAW I

The immediate commentary to Law I is, as far as the
ideas are concerned, its inseperable constituent part; it
reads as follows (/1/, p. 12):

"Lex I. Corpus omne ... mutare.

Projectilia perseverant in motibus suis, nisi quatenus a resistentia aeris retardantur, et vi gravitatis im-
pelluntur deorsum. Trochus, cujus partes cohaerendo perpe-
tuo retrahunt sese a motibus rectilineis, non cessat rota-
ri, nisi quatenus ab aere retardatur. Majora autem Planetarum et Cometarum corpora motus suos et progressivos et
circulares in spatiis minus resistentibus factos conser-
vant diutius.

Lex II. ..."

Translation (/2/, p. 13)
"Law I. Every body ... upon it.

Projectiles continue in their motions, so far as they
are not retarded by the resistance of the air, or impelled
downwards by the force of gravity. A top, whose parts by
their cohesion are continually drawn aside from rectilinear motions, does not cease its rotation, otherwise than as
it is retarded by the air. The greater bodies of the planets and comets, meeting with less resistance in freer
spaces, preserve their motions both progressive and circu-

lar for a much longer time."

The first sentence refers to the _translational_ motion, the second to the _rotary_ one, the third to their _superposition_. Thus, the commentary is consistent with the translation/rotation interpretation closing the preceding section 3, while it is hardly compatible with any of the translations that interprete Law I only for translational motion.

5. NEWTON´S EXPLICIT EMBODIMENT OF THE ROTATION IN LAW 1

The unambiguous enunciation of the embodiment of the rotation in the principle of inertia in Newton´s translation/rotation conception is apparently the result of his previous considerations on the problem of free rotation of an extended body. Herivel has given account of it as follows (/4/, p. 82):

"Free rotation of an extended body

In § 8 of the same MS.V Newton gives a wonderfully just physical appreciation of the free rotation of an extended body. In the first place every such body keeps the same real quantity of circular motion so long as it remains undisturbed. This is the principle of inertia for rotating bodies. Moreover, it continues to rotate about the same axis which always remains parallel to itself provided the endeavours of its four quarters away from the axis of rotation balance. This condition is obviously necessary though apparently not sufficient. But if this balance does not obtain then although there will be a tendency for the body to draw ever nearer to such a balance it will never actually attain to it. And as the actual axis of rotation moves continually in the body, so it will also move continually in space with some kind of ´spiral motion´, always drawing nearer and nearer to a ´centre of parallelism with itself´but never attaining it. ´Nay ´tis

so far from ever keeping parallel to itself that it shall
never be twice in the same position.'

It is to be regretted that Newton did not proceed to
give a quantitative treatment of this problem. It is clear
that his unerring intuition had led him to an almost perfect physical appreciation of the problem."

In a manuscript preceding the Principia Newton formulated a set of six Laws of Motion. Law 1 reads as follows
(/4/, p. 307 Latin, p. 312 English):

"Lex 1. Vi insita corpus omne perseverare in statu
suo quiescendi vel movendi uniformiter in linea recta nisi
quatenus viribus impressis cogitur statum illum mutare.
Motus autem uniformis hic est duplex, progressivus secundum lineam rectam quam corpus centro suo aequabiliter lato
describit et circularis circa axem suum quemvis qui vel
quiescit vel motu uniformi latus semper manet positionibus
suis prioribus parallelus."

"Law 1. By reason of its innate force every body pre-
serves in its state of rest or of moving uniformly in a
straight line unless in so far as it is obliged to change
its state by forces impressed on it. Uniform motion, however, is of two kinds, progressive along a straight line
which the body describes uniformly with its centre, and
circular about a certain axis which either rests or with a
motion of constant size always remains parallel to its
previous position."

Newton's idea, in Law 1 univocally outspoken without
any obscurity, embodies the rotation explicitly in his
translation/rotation conception of the principle of inertia. It operates with the concept of the main axis of inertia of a body, without using this expression, of course.
He deals with it in the manuscript entitled "De Motu Corporum Liber primus", and dated 'Octob. 1684' (/4/, p.321).

"The original text has been subjected to considerable
emendation at certain points, including a number of omis-
sions and insertions. As emended, the manuscript agrees so
closely with the corresponding part of Book I of the Principia[1] that it must represent something very close to, if
not identical with, the final draft of that work, from
which was taken the copy transmitted to the Royal Society
in 1686. ... /4/, p. 321.

[1]Here and elsewhere the reference is to the First Edition."

Significant differences between the original and the
final versions concern, among others, the Definitions.
Here is of interest Definition 2 which in the original
version reads as follows (/4/, Definition 2 p. 321, foot-
note 3 p. 326):

"2.[3] Axis materiae est linea quaevis recta circum
quam materia servato partium situ inter se, in spatio li-
bero absque impedimentis et incitamentis uniformiter re-
volvi possit.

3. Both this and the succeeding definition were omitted
from the Principia. They are of interest for the evidence
they supply of the continuity of Newton's dynamical
thought. See above, Chapter 5, p. 86."

"2.[3] The axis of matter is a certain straight line
about which the matter, its parts maintaining their relative positions, in free space without impediments and in-
citements, can uniformly revolve."

However, the deletion of Definition 2 needn't mean
the abandonment of the translation/rotation conception of
the principle of inertia. The ripened idea may be sus-
tained.

The crucial second sentence in Law 1

"This uniform motion, however, is of two kinds, progressive along a straight line which the body describes uni-

<u>formly with its centre, and circular about a certain axis</u>
<u>which either rests or with a motion of constant size al-</u>
<u>ways remains parallel to its previous position.</u>"

is omitted in the list of Newton's eight various expressions given to the principle of inertia (/4/, p. 30, the
last but one expression). It is, however, paid attention
in another place of Herivel's study; the end of Chapter 5
"The Motion of Extended Bodies" reads as follows (/4/, p.
86):

"Apart from the reference to axes and centre of matter in Definitions 2 and 3 there is no further reference
to rotating bodies in the lectures "de Motu". Nor is there
any indication of any further development of Newton's
thought on this topic in the Principia itself. On the contrary, his erroneous treatment of the precession of the
equinoxes would seem to point to a definite retrogression
in his thought on this subject compared with the original treatment of it in the problem of the collision of
two rotating bodies. There is, however, a very interesting
reference to rotation at the end of his enunciation of the
principle of inertia in MS. Xa: there he continues

Motus autem uniformis hic est duplex, progressivus secundum lineam rectam quam corpus centro suo aequabiliter
lato describit et circularis circa axem suum quemvis qui
vel quiescit vel motu uniformi latus semper manet positionibus suis prioribus parallelus.

It would seem, therefore, that originally Newton had
in mind a principle of inertial <u>rotatory</u> motion besides
that of translatory motion."

The above opinion presents Newton's translation/rota-
tion conception of the principle of inertia as a temporary
matter only. However, in contradistinction to the difficulties of a full quantitative treatment of the general
case of free rotation of an extended body the special case

of a body continuing in the already existing uniform rotation is relatively simple. Therefore, in connection with
the above mentioned evidence of Newton's knowledge, it may
be not surprising that he did not cease thinking of his
translation/rotation conception of the principle of inertia. For to create and work out a profound theory of rotating bodies would be a very demanding task, even though
already Newton's manuscript work on this topic is highly
estimated, e. g. by Herivel in /4/, p. XIII:

"General Introduction. With certain notable exceptions,[1] the whole history of dynamics from Newton to Einstein can be thought of as an exploitation, albeit infinitely ingenious and resourceful, of the definitions,
principles, and propositions in the Principia. ...

[1] For example, the theory of rotating bodies. Curiously,
when Newton came to write Book III of the Principia he
either ignored, or had forgotten, the early, and particularly brilliant work on rotating bodies found in MS. V."

The change-over from Law 1 to Law I consists of two
acts. (i) The commentary on Law I takes over the function
of the second sentence of Law 1, as shown in section 4.
(ii) The univocal term "in linea recta" ("in a straight
line") in Law 1 must be replaced by another one, if the
idea common to both translational and rotary motions is to
be expressed. The interpretation of the term "in directum"
in this function is given in section 3; the following section
6 adds a relevant statement concerning the change-over of
the terms.

6. NEWTON'S CHANGE-OVER FROM UNIVOCAL "IN LINEA RECTA" TO
   EQUIVOCAL "IN DIRECTUM"

Comparing Lex 1 (section 5) and Lex I (section 3),
the differences are not only in leaving out "Vi insita"
and the second sentence in Lex 1, but also in changing
over from "in linea recta" in Lex 1 to "in directum" in

Lex I. It is relevant that in no of his previous enunciations of the principle of inertia Newton used the expression "in directum". His terms of the years 1664-1684 listed in /4/, p. 29-30, are successively (1) in the same
straight line, (2) linea recta, (3) in straight lines, (4)
secundum rectam lineam, (5) in linea recta, (6) in linea
recta, and finally (7) in directum. The last "in linea
recta" appears in Lex 1 quoted in section 5, i. e. with
enunciating the principle still in two sentences, the second being an explicative one. The simultaneity of trans-
ferring the contents of the second sentence of Lex 1 to
the commentary on Lex I and changing over from "in linea
recta" to "in directum" suggests the idea that Newton used
the equivocal term "in directum" intentionally to express
the principle of inertia in his translation/rotation conception. As already mentioned, it must be admitted that
the interpretation of "in directum" for the rotary motion
is, of course, rather strained.

7. NEWTON´S CHANGE-OVER FROM SIX LAWS TO THREE AXIOMS

The structure of the preliminary set of six Laws of
Motion and two Lemmata (/4/, pp. 307-308; MS. Xa.Drafts of
definitions and laws of motion) on the one hand, and the
structure of the final three Axioms, or Laws of Motion with
commentaries and six Corollaries (/1/, pp. 12-19) on the
other hand, demonstrate Newton´s tendency to axiomatization. Law 6, which is a special law of force, namely the
law of resistance of the medium, does not rank among Axioms and is omitted; Law 4 and Law 5 are presented as Corollaries V and IV.

Therefore, having in mind the tendency to axiomatization, it would be rather strange to put the principle of
inertia, understood for the translational motion only, on
the top of the three Axioms, with contents straightfor-

wardly following from Law II. This applies neither in the
case of the full contents of Law 1 nor in the case of its
transformation to Law I with the translation/rotation commentary on it.

The cardinal argument for the translation/rotation
contents of Law I is presented in the next section by Newton's reference to Law I in the case of pure rotation.

8. NEWTON'S REFERENCE TO LAW I IN THE CASE OF PURE ROTATION

The double meaning of the expression "in directum"
used by Newton in the Principia can be unambiguously demonstrated on the following two cases.

(i) The translational motion
/1/, p. 18, Corollarium V, the last sentence:
"... Motus omnes eodem modo se habent in Navi, sive ea
quiescat, sive moveatur uniformiter in directum."
"... All motions happen after the same manner in a ship,
irrespectively of whether it is at rest, or is moving uniformly in the given direction."

In this case, "motus uniformiter in directum" or "a
uniform motion in the given direction" means a uniform motion forwards in the straight line, i. e. a motion with a
constant vector of velocity.

(ii) The rotary motion
/1/, p. 377, Prop. XVII. Theor. XV.
"Planetarum motus diurnos uniformes esse, et librationem
Lunae ex ipsius motu diurno oriri.
Patet per motus Legem I. et Corol.22. Prop. LXVI.
Lib. I. ..."

"Diurnal motions of the planets are uniform; and the li-
bration of the Moon arises from its diurnal motion.
This is obvious from Law I of motion and Corol. 22.
Prop. LXVI, Book I. ..."

Diurnal motions of planets are pure rotations. Considering that Law I is referred to, "motus uniformiter in directum" in Law I, in this case, must mean a uniform rotary motion forwards about an invariable axis, i. e. a motion with a <u>constant vector of angular velocity</u>.

If Law I were meant only for the translational motion, there would be no reason for the above reference.

9. THREE PHASES IN NEWTON'S WORDING OF THE PRINCIPLE OF INERTIA

The final stage of development of Newton's wording of the principle of inertia is the enunciation of the First Law of Motion in a way comprising both translation and rotation, as demonstrated undoubtedly by Newton's reference in the Principia to Law I in the case of rotation of the planets (section 8). The way to this final stage can be followed on the basis of Newton's expressions in his various manuscripts written in the period 1664-1684 and listed in /4/, p. 29-30, as already mentioned in section 6.

Phase 1   Motus in linea recta
            <u>Motion in the straight line</u>

The first six of the above mentioned expressions concern altogether unambiguously the translational motion only. The evidence of it is given by the terms reading or meaning univocally "in the straight line".

Phase 2   Motus progressivus and circularis in linea recta
            <u>Translation and rotation in the straight line</u>

The second phase in Newton's wording the principle of inertia is represented (see section 5) by the expression quoted from the manuscript Xa, according to /4/, p. 321, most probably the last but one prior to the composition of the Principia.

Considering the listed first part of Lex 1 (section
5), the sentence ought to be ranked among the preceding
expressions concerning mere translation. This would be,
however, misleading; on the contrary, Lex 1 is the most
outspoken expression of Newton's translation/rotation conception of the principle of inertia. He combines the expression used up to that time, "in linea recta", with adding another sentence explicitly dealing with the phenomena of rotation and superposition of translation and rotation. Thus, Newton's conception of the principle of inertia is, as to its contents, accomplished. Formally, however, Lex 1 includes a contradiction in terms "circular motion in a straight line". This fact in itself could have
been a sufficient reason for seeking for another true and
exact wording suitable to be published as Lex I in the
Principia.

Phase 3   Motus in directum
          Motion in the given direction

The phase between Lex 1 in the manuscript Xa and
Lex I in the Principia deals with three items.

(i) Axiomatization, i. e. restriction of the number
of laws of motion from six to three. Law 1, not being regarded as a special case of another law, could not be left
out.

(ii) The explanation of the full contents of the
principle of inertia given in Law 1 is presented by the
clear structure of the commentary, appended to Law I,
with these three examples: (1) for the translational motion, (2) for the rotary motion, (3) for the superposition
of both.

(iii) The main problem was how to put both the uniform translational motion and the uniform rotary motion
under one heading, and in such a way as to make it possi-

ble to express the law in one sentence only, similarly
as the other two. The needed expression must include "in
the straight line", however, it must also admit the rotation. The term "in the given direction" fulfils this de-
mand. In Latin with the change-over from "in linea recta"
to "in directum" the problem could have been solved.

It is, of course, clear that neglecting the structure
of the commentary and not respecting the reference to
Law I in the case of the rotation of the planets, the interpretation of Law I just for the translational motion
only appears quite natural even with the novel term "in
directum". To the first translators and to many of their
successors neither the manuscript Xa with two kinds of motion nor the sequence of the terms meaning "in linea recta", interrupted at the critical point by changing over to
"in directum", were known. Also, it is understandable that
the discrepancy between the main sentence of Law I with
the hidden translation/rotation meaning and the commentary
with the open translation/rotation enunciation might have
escaped the translators' attention. The more it can be
understood that they did not notice the sole place in the
whole Principia where Law I is referred to in a clear case
of rotation. Thus, the expression "in a straight line", or
"rectilinear" became common as the univocal equivalent of
"in directum", and, in this sense, it was generally accepted in connection with Law I for so long as the only possible interpretation.

10. A SURVEY OF EVIDENCE FOR THE TRANSLATION/ROTATION
    INTERPRETATION OF LAW I

The following summing-up recapitulates the main characteristics of the arguments dealt with in the precedent
sections.

(1) <u>The main clue:</u>
   The second, explicative sentence in Law 1. (Section 5)
(2) <u>Logical support</u> of the axiomatization:
   The rotation in Law I makes the set of Laws I, II, and
   III irreducible. (Section 7)
(3) <u>Linguistic support</u> of rewording the first law:
   The change-over from "in linea recta" to "in directum"
   takes place in the sequence of Newton's expressions of
   the principle of inertia at the critical moment of rewording Law 1 with the full explicit translation/rota-
   tion commentary. (Section 6)
(4) <u>Internal and external consistence</u> of Law I:
   (i) Correspondence of the kinds of motion considered
      on the one hand in the formulation of Law I, and
      on the other hand in the commentary on it. (Sec-
      tion 4)
   (ii) The explicative second sentence of Law 1 is per-
      fectly projected, point-by-point, to the transla-
      tion/rotation commentary on Law I. (Sections 4, 5)
(5) <u>The cardinal argument:</u>
   The convincing demonstration is given by the reference
   to Law I in connection with the rotary motion of the
   planets. (Section 8)

   Thus, it seems indeed certain that Newton's wording
of Law I should be interpreted in terms of both the
translational and rotary motions. The interpretation of
Newton's wording in the deep-rooted use in the literature
should be, therefore, revised; an example of a veracious
expression of Newton's wording of Law I is given at the
end of Section 3. More details on the topic are given in
/8, 9, 10, 11/.

   As it was stressed already in the introductory Section 1, the proposed revision concerns only the contents

of <u>Newton's wording of Law I</u>; and the question of the formulation and interpretation of the contents of <u>the principle of inertia</u> as seen <u>in contemporary physics</u> is another
problem.

REFERENCES
1. Newton, I., Philosophiae naturalis principia mathematica. Amstaelodami 1723. XXVIII + 484 + VII p.
2. "Sir Isaac Newton's Mathematical Principles of Natural
   Philosophy and his System of the World." Translated into English by Andrew Motte in 1729. The translation revised, and supplied with an historical and explanatory
   appendix, by Florian Cajori. Volume One: "The Motion of
   Bodies." Volume Two: "The System of the World." University of California Press, London 1934, Vol. I - sixth
   printing 1966; Vol. II - seventh printing 1973. 680 p.
3. Thomson, W. and Tait, P. G., "Treatise on Natural Philosophy", Vol. I, Part I. New edition. Cambridge, At
   the University Press, 1879. XVII + 508 p.
4. Herivel, J., "The Background to Newton's Principia.
   A Study of Newton's Dynamical Researches in the Years
   1664-84", Clarendon Press, Oxford 1965. 337 p.
5. Mach, E., "Die Mechanik in ihrer Entwickelung histo-
   risch-kritisch dargestellt", J. A. Barth, Leipzig 1883.
   9. Aufl. 1933. 493 S.
6. More, L. T., "Isaac Newton. A biography", Dover Publications, Inc., New York 1962. An unabridged republica-
   tion of the work first published by Charles Scribner's
   Sons in 1934. 675 p.
7. Šantavý, I., "Newton's First Law", Eur. J. Phys. 7,
   132-133 (1986).
8. Černohorský, M., "Newtonova formulace prvního pohybového zákona. Pokroky matematiky, fyziky a astronomie 20,
   344-349 (1975). ("Newton's Wording of the First Law of

Motion", Advances of Mathematics, Physics and Astronomy.) In Czech.

9. Černohorský, M., "Nová formulace Newtonova prvního pohybového zákona", Folia fac.sci.nat.uni.Purk.Brun. <u>18</u>,
5-36 (1977), Physica 23, opus 1. ("A New Wording of
Newton's First Law of Motion".) In Czech.

10. Černohorský, M., "Problém interpretace Newtonovy formulace prvního pohybového zákona", Folia fac.sci.uni.
Purk.Brun. <u>20</u>, 5-32 (1979), Physica 28, opus 3. ("On
the Interpretation of Newton's Wording of the First
Law of Motion".) In Czech.

11. Černohorský, M.,"Devět Newtonových formulací prvního
pohybového zákona". In: "Pocta Newtonovi." Sborník.
Odborná skupina Pedagogická fyzika FVS JČSMF, Brno
1987. S. 70-86. ("Nine Newton's Enunciations of the
First Law of Motion." In: "In Honour of Newton." Memo-
rial Proceedings of the Physical Section, Union of
Czechoslovak Mathematicians and Physicists, Brno 1987.
Pp. 70-86.) In Czech.

PRIMARILY FORMULATED SCIENTIFIC FOUNDATIONS OF THE ROCKET PROPULSION AND LAUNCHING OF THE ARTIFICIAL SATELLITES IN NEWTON´S "PRINCIPIA"

(On the occasion of the 300 Anniversary of the first publication of "Principia")

Mieczysław Subotowicz

Institute of Physics, M. Curie-Skłodowska University, 20-031 Lublin, Poland

ABSTRACT

In the paper the contribution of I. Newton (1687) to the fundamental laws and fundamental ideas formulation concerning the possibility of the rocket propulsion and that of artificial satellite launching are acknowledged. His book |1| contains the analysis of the motion of planets round the sun as well as that of the projected body round the Earth. The role of the centripetal force plays the gravity force, decreasing as the square of the distance from the centre of the central body. The enclosed illustration was taken from the Newton´s book |1|, as well as all the quotations in the present paper.

1. INTRODUCTION

In 1987 we celebrate the 300th anniversary of the space era. In October 1957 the first artificial satellite of the Earth was launched, the Russian Sputnik. In this connection the names of many important scientists, engineers and technicians contributed to the realization of the first flights of the artificial satellites are mentioned, e.g. K. Ciołkowski, R. Goddard, R. Esnault-Pelterie, H. Oberth, W. Hohmann, A. Szternfeld, W. v. Braun, S. P. Korolew and many others scientists and engineers.

In this connection I would like to present few remarks
on the contribution of I. Newton (1687) to the fundamental
laws and fundamental ideas formulation concerning the possibility and realization of the reactive (jet) propulsion
or rocket propulsion and that of artificial satellites and
space flight. This is also connected with the celebration
of the 300th anniversary of the publication of the distinguished Isaack Newton´s book: "Principia Mathematica
Philosophiae Naturalis", |1|.

2. ROCKET PROPULSION

In the book mentioned above |1| Newton formulated three
principles of dynamics. We shall be interested mainly with
the third principle of dynamics, the well known action-and-
reaction principle. In it is also contained the principle
of the rocket propulsion. As the principle of the momentum
conservation follows from the action-and-reaction principle,
both principles of the description of the rocket flight are
given clearly and precisely by I. Newton for the first time.
At that time (the second half of XVIIth century) rockets
were known mainly as fireworks used at least for about the
last 400 years ago. But nobody analysed the principle of
the rocket motion, scientifically, as the reaction of mass
thrown out of the nozzle of the rocket on the remaining
body and fuel of the rocket. This was first done by I. Newton.

2. HOW TO LAUNCH THE ARTIFICIAL SATELLITE
OF THE CELESTIAL BODY?

An important part of Newton´s book |1| contains the
analysis of the motion of planets round the sun. I would
like to keep my attention on the second book and its im-

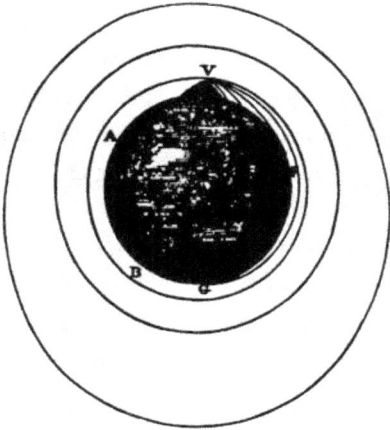

Fig. 1. This figure taken
from Newton´s book |1|
proves the possibility of
launching the artificial
satellite.

a less velocity describes the lesser the arc VD, and with
a greater velocity the greater the arc VE, and augmenting
the velocity, it goes further and further to F and G, if
the velocity was still more and more augmented, it would
reach at last quite beyond the circumference of the earth,
and return to the mountain from which it was projected".
... "Its velocity, when it returns to the mountain, will be
no less than it was at first; and retaining the same velocity, it will describe the same curve over and over, by the
same law".
... "But if we now imagine bodies to be projected in the
directions of lines parallel to the horizon from greater
heights, as of 5, 10, 100, 1000, or more miles, or rather as
many semidiameters of the earth, those bodies, according to
their velocity, and the different force of gravity in different heights, will describe arcs either concentric with
the earth, or variously eccentric and go on revolving through
the heavens in those orbits just as the planets do in their
orbits".
    As we see, in the Newton´s text, it contains an exact
indication on the possibility to establish artificial satel-

portant part: "The system of the world". After a short discussion of the history of the growing understanding of the
structure of the solar planetary system by Chaldeans, Egyp-
tians, Greeks and Romans, the book deals with the principle
of circular motion in free space. The force responsible for
the circular motion of the celestial bodies is called "a
centripetal force, as it is a force which is directed towards some centre; and as it regards more particularly a
body in that centre, we call it circumsolar, circumterres-
trial, circumjovial; and so in respect of other central
bodies".

To explain how to understand "that by means of centripetal forces the planets may be retained in certain orbits
... we consider the motions of projectiles". ... "For a
stone that is projected by the pressure of its own weight
forced out of the rectilinear path, which by the initial
alone it should have pursued, and made to describe a curved
line in the air; and through that crooked way is at last
brought down to the ground; and the greater the velocity is
with which it is projected, the further it goes before it
falls to the earth. We may therefore suppose the velocity
to be so increased, that it would describe on arc 1, 2, 5,
10, 1000 miles before it arrived at the earth, until at
least, exceeding the limits of the earth, it should pass
into space without touching it". This is the first scientific formulation of the possibility how to launch artificial
satellites of the Earth. I. Newton accepts that "there is
no air about the earth, or at least it is endowed with
little or no power of resisting".

In Fig. 1"AFB represents the surface of the earth, C
its centre VD, VE, VF the curved lines which a body would
describe if projected in and horizontal direction from the
top of an mountain successively with more and more velocity" ..."for the same reasons the body projected with

lites of the earth, moving along the closed trajectories, circular or elliptic ones (eccentric orbits). The gravity force determines the trajectory of the body. This force plays the role of the centripetal force, necessary to move the body around the central (large) body. This centripetal force is directed to the centre of the central body. It decreases inversely as the square of the distance from the centre of the central body earth (earth, sun, Jupiter or other planets).

REFERENCES

|1| Isaac Newton: Mathematical Principles of Natural Philosophy and His System of the World, translation into English by Andrew Motte in 1729, revised by Florian Cajori, vol. 2: The System of the World, University of California Press, Berkeley and Los Angeles, 1962 (fifth printing).

FROM ISAAC NEWTON TO ALBERT EINSTEIN

Jerzy Rayski

Institute of Physics, Jagiellonian University,
Reymonta 4, 30-059 Kraków, Poland

ABSTRACT

The concept of inertial coordinate systems in Newton´s
and in Einstein´s theories is discussed and quasi-
-inertial coordinate systems are defined even in General Relativity.

1. INTRODUCTION

The main fields of interest and of research of the two
greatest physicists in the history of mankind exhibit striking analogies: Both formulated equations of motion for
point particles, both solved the problem of gravitational
forces, to the extent permitted by the contemporary means
and possibilities, and both achieved a unification of apparently quite different scales and research domains: terrestrial physics and cosmology. However, whereas Newtonian
cosmology was limited to the planetary motion around the
Sun, Einsteinian cosmology involves the whole Universe.
Last not least, both were deeply involved into optical pro-
blematics and both represented a corpuscular point of view
upon the nature of light.

2. INERTIAL COORDINATE SYSTEMS

In my talk I shall concentrate attention upon the concept of inertial frames of reference and inertial coordinate systems fixed to the reference bodies, i.e. in which
the latter are resting. Einstein's concept of inertial coordinates does not differ much from that of Newton as far as
Special Relativity (SR) is concerned. The only difference is
that a transition from one to another system of coordinates
is governed by the Lorentz group but not by the Galilei group
of transformations. On the other hand, the difference between Special and General Relativity (the latter admitting
a general group of transformations) is essential since - according to Einstein - no inertial frames of reference exist
in General Relativity (GR) at all, all coordinate systems
are on equal footing and no one is privileged.

If this statement were true, we would be in trouble
with securing correspondence between GR and earlier theories. Some authors point out that correspondence is ensured
because it is always possible to introduce locally geodesic
coordinates, but it does not seem to be sufficient. Obviously, it is not possible to introduce a Cartesian and inertial
coordinate system (whose coordinate axes are straight lines
orthogonal to each other) since straight lines do not exist
in curved spaces (unless there is a symmetry) but question
arises whether it is possible to define an analogue differing
as little as possible from the inertial one. I shall try to
convince you that such a very natural possibility exists, at
least if the gravitational field is not too violent and
quickly changeable but satisfies some condition of stability.
Such situation might be called gravitationally stationary.

3. CONSERVATION LAWS

If a covariant divergence of a vector field j^{μ} vanishes

$$\nabla_{\mu} j^{\mu} = 0 , \tag{1}$$

where ∇_{μ} denotes a covariant derivative, then it is equivalent to a continuity equation for the density

$$\partial_{\mu} (\sqrt{-g} \, j^{\mu}) = 0 , \tag{2}$$

where g is the determinant of the metric tensor components

$$g = \det g_{\mu\nu} . \tag{3}$$

The equivalence of (1) and (2) follows from the fact that a covariant derivative of a vector field is

$$\nabla_{\mu} j^{\mu} = \partial_{\mu} j^{\mu} + \Gamma^{\rho}_{\rho\mu} j^{\mu} , \tag{4}$$

where

$$\Gamma^{\rho}_{\rho\mu} = \frac{1}{\sqrt{-g}} \, \partial_{\mu} \sqrt{-g} . \tag{5}$$

The energy-momentum-stress tensor T^{μ}_{ν} representing the sources of the gravitational field and appearing to the right-hand side of Einstein's equations

$$\frac{1}{\varkappa} G^{\mu}_{\nu} = T^{\mu}_{\nu} \tag{6}$$

also satisfies the equation

$$\nabla_{\mu} T^{\mu}_{\nu} = 0 , \tag{7}$$

but unlike vector fields where from a vanishing of a covariant divergence (1) there follows the continuity equation (2) analogous energy-momentum conservation laws in the form of continuity equations do not follow from (7) due to the appearance of the additional third term in the formula for a covariant divergence of tensors

$$\nabla_\mu T^\mu = \partial_\mu T^\mu + \Gamma^\rho_{\rho\mu} T^\mu_\nu - \Gamma^\mu_{\nu\rho} T^\mu_\mu .$$

(8)

This fact has been interpreted so that the energy-momentum of the substantial sources of gravity (different fields and particles) do not constitute yet the whole energy-momentum, since the free gravitational field itself is also a seat and carrier of energy-momentum giving rise to an additional term τ^μ_ν constructed with the help of first derivatives of the metric tensor components. Einstein constructed this term so that

$$\partial_\mu [\sqrt{-g}(T^\mu_\nu + \tau^\mu_\nu)] = 0 ,$$

(9)

but the trouble was that it does not transform like a tensor under the general coordinate transformations. Therefore Einstein called it a "pseudo-tensor" of the energy-momentum of the gravitational field. The fact that energy-momentum are not well defined and properly localized quantities may be seen best from the fact that τ^μ_ν involve only first derivatives of the metric tensor and may be brought to vanish at an arbitrarily chosen point by going over to a locally geodesic system of coordinates. Thus, τ^μ_ν cannot be a genuine tensor and something seems to be wrong with general relativity.

4. CONSERVATION LAWS AND INERTIA

The troubles with the conservation laws would disappear if the last term in (8) were absent. Question arises in which cases and under what conditions this last term could disappear:

$$T_\nu \overset{=}{def} \Gamma_{\nu\varsigma}^{\ \mu}\ T_\mu^\varsigma \overset{?}{=} 0 \tag{10}$$

Vanishing of this term would obviously mean that energy-
-momentum of the material (substantial) sources alone were
conserved, or - in other words - that there is no transfer
of energy-momentum between the sources and the free gravitational field. This is possible provided the material sources
are so to say "gravitationally stationary", i.e. neither ab-
sorption nor emission nor scattering of gravitational wave
(gravitons?) by matter fields occurs. To be more precise:
a vanishing of the term T_ν defined by (10) is a condition
upon the system of coordinates which may be satisfied if
the physical system under consideration is gravitationally
stationary. Such coordinate system will be called inertial
(ex definitione).

    Obviously, the condition (10) will be not satisfied
in arbitrary coordinates, even in the case of a stationary
system, but such a "violation" of conservation laws for
energy-momentum of the sources alone is then interpretable
merely as a coordinate effect.

    In this way it could be shown that some coordinate
system deserve the name "inertial" and are very natural
even in curved spacetimes, provided the sources of gravity
satisfy some stability or stationarity conditions. However,
even if such conditions are not fulfilled we might still
try to introduce coordinates that are as similar to inertial ones as possible, i.e. such that the left-hand side of
(10) were as small as possible. Such requirement is just a
condition upon a coordinate system to be called "quasi-
-inertial".

5. DIFFICULTIES WITH A CANONICAL FORMULATION

Hilbert has shown that Einstein's equations (6) follow as Euler-Lagrange equations from a Lagrangian differing from a scalar curvature by a divergence and involving only first derivatives of the metric tensor. Thus, it is a theory of a Lagrangian form, but it is not of the usual canonical form, due to the appearance of constraints. The equations defining the four canonically cojugate momenta

$$\pi^{\mu o} = \frac{\partial \mathcal{L}}{\partial \dot{g}_{\mu o}} \qquad (11)$$

are not soluble for $\dot{g}_{\mu o}$ which follows from the fact that the Lagrangian does not involve these four time derivatives quadratically as it used to be for tensor fields describing bosons (if quantized). A peculiar role of some metric tensor components follows also from the fact that due to the Bianchi identities the field equations are not independent of each other. Four among $g_{\mu\nu}$ may be chosen freely, i.e. settled by means of arbitrary coordinate conditions (being at the same time gauge fixing conditions). The troubles with a usual canonical formulation, or how to deal with the peculiar constraints of general relativity may be also regarded as responsible for the unsurmountable difficulties with quantization of the gravitational field.

6. A POSSIBLE SOLUTION

The difficulties connected with the problems of existence of inertial (or quasi-inertial) coordinate systems and with a truly canonical formulation may be solved simultaneously. One has only to add to the usual Lagrangian of gravitational field theory an additional term

$$\mathcal{L} \longrightarrow \mathcal{L} + \mathcal{L}', \qquad (12)$$

58

where

$$\mathcal{L}' = \varepsilon \, T_\mu \, g^{\mu\nu} T_\nu \qquad (12')$$

with T_ν being the left side of (10). This new Lagrangian is no more generally covariant but, instead, it involves all time derivatives $\dot{g}_\mu$ quadratically so that the new field equations

$$G^\mu_\nu + H^\mu_\nu = \varkappa \, T^\mu_\nu \qquad (10')$$

become second order field equations, all of them on equal footing, and destroying the Bianchi identities.

Thus, the new equations determine simultaneously all ten metric tensor components, i.e. not only the metric of spacetime but also a privileged system of coordinates. Since H^μ_ν vanished together with T_ν the new field equations comprise all the usual Einsteinian solutions that describe gravitationally stable (or stationary) situations but expressed already in some inertial coordinate systems. Beyond such solutions the new equations involve some extra non-stationary solutions which are no more solutions of the original equations of Einstein (10), but for which the extra term H^μ_ν cannpt be brought to vanish by any choice of coordinates. These solutions also determine not only the metric of spacetime but also a system of coordinates that might be regarded just as a natural extrapolation and continuation of the concept of inertial coordinate systems.

It should be noticed that the new terms appearing in the Lagrangian and in field equations and violating the general covariance vanish in domains devoid of matter, i.e. in domains where the tensor T^μ_ν vanishes. It means that empty space remains generally covariant in the broad Einsteinian sense of this word. Some violation of general covariance might occur only inside matter. This seems intelligible since matter might constitute a substitute for

the cosmical aether, might be defining and preferring a
certain class of frames of reference.

Just, Einstein's thesis that all systems of coordinates
are on equal footing should be paraphrased as follows:

"All systems of coordinates are equal but some are
more equal than the others".

A P P E N D I X

The philosophers of earlier epochs tried to construct
great philosophical systems, to produce synthetic pictures
of the Universe, its mechanism and cosmology, as well as
to formulate a "Weltanschauung". In view of the progress of
modern science their endeavours must appear as primitive
and extremly naive.

Disappointed by such premature attempts modern philosophers try to be more modest. Instead of formulating great
systems they practise what is called an analytic philosophy
based chiefly upon the progress of modern logic and mathematics.

However, a revolutionary progress of exact sciences,
especially of physics that is wittnessed in the XX-th Century, especially a formulation of the theory of relativity
and of quantum mechanics opens new possibilities and new
horizons also for philosophy. Especially the progress of
our knowledge about the most elementary constituents of
matter and their fundamental interactions has led in the
last decade to a new understanding of the laws of Nature,
their symmetries and supersymmetries and to some unforseen
and quite unexpected relations between apparently different
types of phenomena as well as to discoveries of some intimate relations between events on a micro-scale with those on
the scale of the whole Universe.

In this new situation physicists are making successful
attempts to formulate what they call "a unified theory" or

somewhat humoristically "a theory of everything" (TOE). One should not forget, however, that the first attempt towards such unification was done by Newton (the anniversary of whose PRINCIPIA we are just celebrating) who unified the two apparently quite different types of phenomena: terrestrial with celestial gravity.

Only nowadays there appear good chances to supplement and to extend the analytic philosophy by some more synthetic research and to try to formulate an integrating philosophy and a synthetic system, this time based upon a solid and more sound basis of exact sciences.

THE CORRESPONDENCE BETWEEN THE EINSTENIAN AND NEWTONIAN THEORIES OF GRAVITATION

T. Grabińska, M. Zabierowski
Technical University of Wrocław, Wrocław

... który jesteś - Boże,
Ja także jestem... C. Norwid

ABSTRACT

The Newtonian theory of gravitation as a special case of general relativity is studied. The correspondence relation between two theories is tried to be recorded. Its usually considered form, i.e. the limiting transition between respective formulae cannot be longer accepted as a full correspondence. Similarly the correspondence between automorphism groups is not sufficient. The correspondence between appropriate gauge theories of gravitation is postulated.

1. INTRODUCTION

We shall consider the correspondence relation between two theories of gravitation i.e. the Newtonian theory /NT/ and the general relativity /GR/. Their kinematic counterparts are the Galilean kinematics /GK/ and the Einsteinian special relativity /SR/, respectively. Because of different objects of the two pairs of theories one ought not to think that the correspondence relation between two kinematic theories transfers itself to the second pair automatically.

In the present paper we take into account the two step formulation of the correspondence relation[1]. It suggests that more general physical theory should fulfil the following two requirements in respect to the simpler one: 1° the cumulativity of knowledge and 2° the formal correspondence between formalisms of the theories. It is easy to show that SR and GK satisfy both postulates. The cumulativity of knowledge is realized in SR through the empirical consistency of it with the narrower theory /i.e.GK/ in their common range of applicability. The correspondence between the mathematical models of both theories /i.e. Minkowski's and Galileo's space-time, respectively/ is also very impressive. It is carried out through making a transition from the automorphism group of Minkowski's space-time /i.e. the Lorentz's group of transformations/ to the Galileo's group of transformations. That is the case when $v/c \longrightarrow 0$ [2], where v means the velocity of rectlinear and uniform motion, c stands for the light velocity constant.

The Riemannian space-time is the mathematical model of GR and from the very definition it is invariant in respect to every transformation of space-time coordinates. Hence the correspondence between the global symmetry of GR space-time model and the respective global symmetry of NT cannot be defined in a unique way. There exists indeed a possibility to look for the correspondence between local /gauge/ symmetries of both theories. However the question of local symmetries of GR remains still open. So does the correspondence between GR and NT.

2. THE NEWTONIAN APPROXIMATION IN THE GENERAL RELATIVITY

The fundamental equations of GR are given as follows

$$R_{\mu\nu} - \frac{1}{2} g_{\mu\nu} R = \kappa T_{\mu\nu}, \quad \mu, \nu = 0,1,2,3 \quad /1/$$

where $R_{\mu\nu} - \frac{1}{2} g_{\mu\nu} R = G_{\mu\nu}$ is called the Einsteinian tensor
and expresses geometrical properties of the curved space-
time; $T_{\mu\nu}$ is the stress-energy tensor which reflects the
distribution of matter and fields in the space-time, κ is
a constant. In a general sense the left-hand side corres-
ponds to the gravitational potential whereas the right-
hand side accords with the source of gravitational field.
The eqs./1/ multiplied on both sides by the metric tensor
$g_{\mu\nu}/x/$ imply

$$R = -\kappa T. \quad /2/$$

The insertion of /2/ back to /1/ gives the new form of
the GR gravitational equations

$$R_{\mu\nu} = \kappa \left(T_{\mu\nu} - \frac{1}{2} g_{\mu\nu} T \right) . \quad /3/$$

In the case of weak gravitational field the velocities v
of bodies are negligible if compared with c, $v \ll c$. Hence
there remains only one non-zero stress-energy tensor component $T_{oo} = c^2 \varrho$ and eqs./3/ are reduced to only one

$$R_{oo} - \frac{1}{2} g_{oo} R = \kappa c^2 \varrho . \quad /4/$$

In the weak gravitational field approximation eq./4/ sets
finally into the following relation

$$R_{oo} \approx \kappa c^2 \varrho/2 ,$$

where as before ϱ denotes the matter density.

On the other hand in the approximation of static gravitational field R_{oo} is given by

$$c^2 \Delta \left(g_{oo}/2 \right) \approx \kappa c^4 \varrho/2, \quad /5/$$

where Δ stands for the Laplace´s operator.

In the Newtonian theory of gravitation the fundamental law for the gravitational potential φ is expressed
by the Poisson equation

$$\Delta \varphi = 4\pi k \varrho , \quad /6/$$

where k means the gravitational constant. The comparison
of formula /5/ and /6/ leads to an identification of the

Newtonian gravitational quantities φ and k with the
Einsteinian g_{oo} and $\varkappa$ in the weak static field approximation

$$g_{oo} = 2\varphi/c^2 + const, \qquad \varkappa = 8\pi k/c^4 . \qquad /7/$$

Formula /7/ can be also rewritten in the form

$$g_{oo} = \left(1 + 2\varphi/c^2\right) + const , \qquad /8/$$

where const=0 because the Newtonian gravitational field
has to vanish in the points "distant" from its source.

The last formula suggests that the remaining compo-
nents of $g_{\mu\nu}$ be of the following form: $g_{ij} = \delta_{ij}$, $g_{oi} = 0$,
i,j=1,2,3. However from the careful considerations of this
approximation it is concluded that the question is more
complex because one obtains [3]

$$g_{\mu\nu} = \eta_{\mu\nu} + h_{\mu\nu} , \qquad /9/$$

where $\eta_{\mu\nu}$ is the Minkowski's tensor and $h_{\mu\nu}$ values are
of the range $\sim \left|2\varphi/ c^2\right|$.

3. THE DISCUSSION ABOUT THE CORRESPONDENCE RELATION BETWEEN GR AND NT

We shall examine the reduction of GR to NT which is
presented in the previous section. The correspondence relation between two theories is understood here as 1°- a
relation between their empirical contents /cumulativity
requirement/ and 2°- their formal consistency in the common range of application /limiting transition requirement/.
It is claimed [4] that theory T´satisfies the cumulativity
postulate in respect to theory T iff the empirical content
of T is represented /recorded/ in T´with a comparable precision. This requirement is fulfilled for T´=:GR and T=:NT.
NT describes weak gravitational field which can cause only
the slow motions /v≪c/.In the empirical reality of NT the
GR-theory reflects all the phenomena with the precision
comparable to the description within the framework of NT.
In the common range of application numerical values of GR-

physical quantities differ from the Newtonian respective ones only in powers of term c^{-2}.Such terms are negligible in the Newtonian approximation.

The second postulate of correspondence between T' and T requires deriving T from T'supplied with additional assumptions which are not contradictory to T'.This postulate concerns the formalism of both theories and in fact ammounts to a limiting transition between the respective formulae of the theories. In the case of GR and NT the transition can be realized in two ways. Firstly,the individual formulae of GR turn into the NT-formulae under the assumption of weak and static gravitational field,as e.g. in section 2. Secondly,since GR and NT are the theories which can be formulated in the uniform mathematical language of Riemannian geometry[5]there should be required the limiting transition between the mathematical models of the theories, e.g. between their groups of automorphisms. In connection with the last meaning of the limiting transition there appear however some important questions unresolved, yet.

4. THE UNRESOLVED PROBLEMS OF CORRESPONDENCE RELATION BETWEEN THE GROUPS OF LOCAL AUTOMORPHISM OF THEORIES

Both theories /GR,NT/,under consideration here,are dynamic ones.Their mathematical models are given by the respective curved spaces. Only locally it is possible to introduce an affine space into the space-times of GR and NT.The result of the limiting transition between the affine spaces is well known:the Lorentz group of automorphisms of the Minkowski space transforms   into the Galilean group under the assumption that $v/c \rightarrow 0$.This transition between kinematic theories is not of the consequence for the dynamic theories under examination.

On the other hand the correspondence between the global space-time automorphisms of GR and NT is impossible

66

because of a priori non-existence of any specific automorphism group of the GR space-time. There remain only
investigations of local /gauge/ automorphisms of both
theories. In this way the methodological searches in the
field of the correspondence relation and physical inquires
about a gauge theory of gravitation [6] approach each other
with respect to their goals. Unfortunately, there are not
still any ultimate results in the field of gauge formulation of gravitational theory and therefore the correspondence relation between GR and NT is not fully recognized.

The considered problem situation allows to state
the following:

1. Amsterdamski is right when he claims [7] that the cumulativity requirement is weaker than the postulate of
   formal correspondence.The absence of cumulativity of
   more general theory in respect to the narrower one im-
   plies the absence of formal correspondence between them;
   the confirmation of the cumulativity /as in the case of
   GR and NT/ is not decisive for formal correspondence
   relation.

2. The correspondence relation, understood in the way as in
   the present paper,can be regarded not merely as a descriptive tool but also as a normative factor for the
   formal structure of the very theories. In the case of
   GR the normative function of the correspondence relation
   appeared in the choice of tensor $G_{\mu\nu}$. On account of mathematical conditions the form of left-hand side of /1/
   could be more general but the Newtonian approximation
   /which was treated as a realisation of correspondence/
   required just $G_{\mu\nu}$. Thus, the formulation of GR was also
   conducted by the correspondence relation.

3. In the case of the correspondence between the automorphism groups of GR and NT the formal correspondence postulate demands further searches for the gauge theory of

gravitation or a formulation of both theories in more
advanced mathematical language.

4.The form of correspondence relation between theories is
closely connected with the formalism of the theories
under consideration. In the case of advanced mathematical models of the theories the requirement of limiting
transition between their automorphism groups seems to
be reasonable and promising not only for methodology
but also for an internal structure of the theories.
Similarly to the case of simpler form of correspondence
relation, i.e. the limiting transition between the
separate formulae of theories, the correspondence
between the automorphism groups is perhaps not the last
word concerning the form of correspondence.

REFERENCES

[1] Grabińska,T., Zabierowski,M.,"Orelacji korespondencji
między teoriami Einsteina i Newtona", Z Zagadnień
Filozofii Przyrodoznawstwa i Filozofii Przyrody,VI,145-
153, 1984.

[2] Cf discussion about the formal correspondence between
SR and GK in Zahar, E.,"Why did Einstein´s programme
supersede Lorentz´s?", in: Method and Appraisal in the
Physical Sciences, ed. C. Howson, Cambridge University
Press, Cambridge, pp.257-259, 1976.

[3] i.e. Pauli, W., Theory of Relativity, Pergamon Press,
London, Ch.4, 1958.

[4] Suszko, R.,"Formal logic and the development of knowledge", in: Problems in the Philosophy of Science,3,
Amsterdam,pp.210-222, pp.227-230, 1968.

[5] Misner, C.W. et al., Gravitation, W.H. Freeman and Co.,
San Francisco, 1973.

[6] Cf Abers, E.S., Lee, B.W., "Gauge Theories", Physics
Reports 9,1, 1973; Utiyama, R.,"Invariant Theoretical

Interpretation of Interactions", Physical Review 100,
1597, 1956; Kibble,T.W., "Lorentz Invariant and the
Gravitational Field", Journal of Mathematical Physics
2, 212, 1961; Hehl. F,W,, "Four Lectures on Poincaré
Gauge Theory", Preprint ORO 3992/380 - Texas University.
Austin, 1979.
7 Amsterdamski, S., Między doświadczeniem a metafizyką,
Warszawa, pp. 208-209, 1973.

VELOCITY OF LIGHT

M. SUFFCZYŃSKI

Institute of Phycis, Polish Academy of Sciences
Al. Lotników 32, Warsaw 02-668, Poland

ABSTRACT

Progress in our knowledge of the velocity of light and of the gravitational constant since the time of Newton is briefly described. While the value of the velocity of light is one of the best determined among the fundamental constants of physics and leads to the definition of the metre in terms of the second the gravitational constant is the least precisely known of the fundamental constants.

Two of the fundamental constants of physics are known since the time of I. Newton: the velocity of light and the gravitational constant. The transit time across the orbit of the earth due to finite velocity of light was determined by O. Roemer in 1676 from the difference in time of the eclipses of the moon Io of Jupiter |1|. In 1728 J. Bradley explained the aberration of star light by the ratio of the speed of the earth around the sun to the speed of light and determined the value of the constant of aberration. H. L. Fizeau (1849) and L. Foucault (1862) measured the velocity of light in terrestrial experiments |2|. A. A. Michelson in his measurements determined the value of the velocity of light in vacuum c = 299796 km/s with a limit of error ₊4 km/s. Recent determinations of the velocity of light based on simultaneous high-precision determinations of the frequency and wavelength of highly monochromatic stabilized laser radiation sources |3,4| have reduced the uncertainty down to 0.0035 ppm (parts per million) and led the Comité Consulatif pour la Définition du Metre of the Comité Inter-

national des Poids et Mesures to the adoption to the value c = 299792458
m/s as an exact defined value |5|. Since 1983 the metre is defined as
the length of path travelled by light in vacuum during a time interval
1/299792458 of a second |6|. The second is the duration of 9192631770
periods of the radiation corresponding to the transition between two
hyperfine levels F = 4 - F = 3 of the ground state of the cesium-133
atom.

The gravitational constant G has been measured first by
H. Cavendish in 1798 with a torsion balance. The oscillating torsion
balance measurements at the National Bureau of Standards have been
performed with improvements by Heyl and Chrzanowski (1942) |7|. The
weighted mean of their results was adopted for the recommended value,
$G = 6.6720(41) \ 10^{-11} \ m^3kg^{-1}s^{-2}$ (615 ppm), in the 1973 adjustment of
the fundamental constants |3|. The measurements have been improved by
Luther and Towler (1982) |8|. Their result was adopted as the recom-
mended value, $G = 6.67259(85) \ 10^{-11} \ m^3kg^{-1}s^{-2}$ (128 ppm), in the 1986
adjustment of the fundamental constants |5|.

The velocity of light "perhaps nature´s most universal physical
constant" |9| has been determined with the highest precision among the
fundamental constants. The gravitational constant is also universal
but there is no well established relationship between it and other
physical quantities. It stands uncoupled from the remainder of the
adjustment |6|. It remains determined with the largest assigned un-
certainty among the fundamental constants of the 1986 adjustment.

REFERENCES

1. Wróblewski, A., Am. J. Phys. 53, 621 (1985).
2. Whittaker, E., A History of the Theories of Aether and Electricity,
   T. Nelson and Sons Ltd., London 1951.
3. Cohen, E.R. and Taylor, B.N., J. Phys. Chem. Ref. Data 2, 663 (1973).
4. Mulligan, J.F., Am. J. Phys. 44, 960 (1976).
5. Cohen, E.R. and Taylor. B.N., The 1986 Adjustment of the Fundamental
   Physical Constants, CODATA Bulletin No. 63, November 1986, Pergamon
   Press.

6. Cohen, E.R., Fundamental Physical Constants, in: Gravitation Measurements, Fundamental Metrology and Constants, Erice, Sicily 1987.
7. Heyl, P.R. and Chrzanowski, P., J. Res. Nat. Bur. Stand. $\underline{29}$, 1 (1942).
8. Luther, G.G. and Towler, W.R., Phys. Rev. Lett. $\underline{48}$, 121 (1982).
9. Terrien, J., Rep. Prog. Phys. $\underline{39}$, 1067 (1976).

NEWTON AND THE MATHEMATICAL CONCEPT OF SPACE

R. DUDA

Institute of Mathematics
University of Wrocław
POLAND

ABSTRACT

Greeks did not possess the mathematical concept
of a space and their straight lines were seg-
ments. An infinite straight line appeared with
the Principle of Inertia and it was Newton who
has drawn full consequences of it by introducing
the concept of an absolute space in which all
movements could be described. He meant by it
the only then possible candidate, that is, the
space underlying Euclidean geometry. After being
named, however, the concept of a space has begun
to live on its own and the paper recalls that
story in XIX and XX centuries.

One of the most honoured parts of the Greek intellectual heritage is geometry. The word itself means measuring
the Earth but with the Euclid's "Elements" (III century
B.C.) it became the name of a mathematical theory of a
great convincing power and beauty. Reasoning "more geome-
trico" has become since the ideal of a scientific procedure and the Euclidean theory has been promoted to the role
of a model of an exact science.

Although the theory refers to points, plane figures,
bodies and the like, nowhere in the book of Euclid neither
in any other mathematical treatise of antiquity there appears a word meaning space as we understand it. Strange
as it sounds, Greeks did not even possess such a word,

to say nothing of a mathematical concept of a space. Their
"kosmos" meant rather integrity and order, later the Universe, "chora" was closely related to the Platonic philosophy of reflection, and "physis" meant Nature. Also Latin
"spatium" denoted in that time place or distance and did
not possess the meaning which Newton gave to it in "Principia" and which we accepted since then (cf. <12>).

Change came with the Renaissance emergence of the modern science. One of its characteristic features was the
role of mathematics which, in contrast to the prevailing
anty-empiric attitude of ancient Greeks, became now the
language of the Nature. For instance, Descartes (1596-
-1650) proclaimed that he "neither admits nor hopes for
any principles in Physics other that those which are in
Geometry or in abstract Mathematics, because thus all
phenomena of Nature are explained and some demonstrations
of them can be given" (cf. <4>, p. 325). Thus the essence
of science becomes mathematics and the objective world is
geometry incarnate.

However, it was Galileo (1564-1642) who became, more
than any other man, the founder of methodology of the modern science. His famous manifesto of 1623 reads (quoted
after <4>, p. 328-329): "Philosophy <nature> is written
in that great book which ever lies open before our eyes -
I mean the Universe - but we cannot understand it if we
do not first learn the language in which it is written.
The book is written in the mathematical language, and the
symbols are triangles, circles and other geometrical figures, without whose help it is impossible to comprehend
a single word of it; without which one wanders in vain
through a dark labirynth."

Galileo's program has been fully accepted and put to
work by Isaac Newton (1642-1727 . Like Galileo, Newton has
been far more engrossed in science than in mathematics.
Unlike Galileo, however, Newton has also greatly develop-

ped mathematics itself.

Undoubtedly, Newton's greatest mathematical achievement is building the Calculus. To be precise, he was not
the first to work in this area (cf. <1>), but he was the
man whose imagination and audacity have allowed him to
make some decisive steps in generalizing ideas advanced
by many men before, to establish full-fledged methods,
and to show their power in dealing with some fundamental
problems in mathematics and science.

The most influential publication involving his Calculus were "Principia Mathematica Philosophiae Naturalis"-
<6>. Although the book is devoted to mechanics, it has
enormous importance also for the history of mathematics
proper, not only because it presents Newton's powerful approach to the Calculus and shows its power, but rather because it deals with the entirely new topics and initiates
new paths that were to be explored during the next two
hundred years in the course of which an enormous amount
of Analysis and geometry has been created.

At the beginning of his "Principia" Newton states his
three famous "Axioms, or Laws of Motion". The first law is
the Principle of Inertia and reads: Every body continues
in its state of rest, or of uniform motion in a right
<straight> line, unless it is compelled to change that
state by forces impressed upon it (<6>, vol. I, p. 13).

Although the Principle of Inertia has been nurtured by
some men before Newton, e.g. by Galileo and Descartes, it
was Newton who gave to it the final form and drew important
consequences. First of all, the Principle of Inertia requires the notion of an infinite straight line and this is,
to the best of my knowledge, the first time that there appeared explicitly, with a great persuasive force, an infinite, in both directions, straight line. Importance of that
notion lies in the fact, which Newton saw clearly, that

such an infinite straight line requires a place to put in, with some system of reference with respect to which it is a straight line and not something else. Newton answered by introducing the general concept of an absolute space in which all motions take place and can be described by his dynamics.

As one can see from his writings, the absolute space was for Newton like a laboratory without walls or a box without faces, a sort of a container comprising all natural bodies and yielding place to all natural phenomena, but notsustaining any influence of theirs. Its properties do not depend on what it contains. The absolute space exists in itself and is not obliged in its existence to anything in the world. It remains always and everywhere the same and invariant. Its all points are equivariant and equal, and so are all directions. It has no boundaries, extends infinitely in all directions, and has infinite volume.

However obscure such an idea may appear to be, it had an immediate mathematical appeal. And within reach there was a ready model, namely Euclid´s geometry based upon Pythagorean metric (the underlying set of points of that geometry has only now been recognized as an independent mathematical concept) and supplied with the handy tools developped by Descartes in his "La Gèometrie" from 1673.

Newton did not provide mathematical description of his absolute space, but whenever he had to resort to it, it was quite clearly the space of Euclidean geometry. In that way the absolute space of Newton has been identified as the three-dimensional Euclidean space E^3, and that space remained a cornerstone of physics and mathematics for at least the next two centuries.

Once the concept of a space has been introduced, it could become an object of a study, starting a life on its own. The process begun in XVIII century with the contestation of the famous Vth postulate of the Euclid´s "Elements". With the help of the notion of an infinite straight line John Playfair (1748-1819) has formulated it in 1795 in the now common form:

V. Through a given point P not on a line l, there is only one line in the plane of P and l which does not meet l.

The problem was: Whether this axiom is necessary? In other words, can it be proved on the basis of the other axioms (and in this way disposed of) or can it be replaced by another one, but simpler and more intuitive.

After several decades there came the discovery, made independently by Carl Gauss (1777-1855), Nicolai Lobatchewsky (1793-4856), and Janos Bolyai (1802-1860), that there is another geometry for the space underlying E^3, a non-euclidean one, in which the Vth postulate has been replaced by the following one.

V´. Through a given point not on a line l, there are at least two lines in the plane of P and l which do not meet l.

Except for the Vth postulate replaced by that V´th, all the other axioms and postulates of the Euclidean geometry remain valid and this new geometry is called nowadays hyperbolic, denoted H^3. And soon came the discovery, due to Bernhard Riemann (1826-4866), that there is yet another non-euclidean geometry for the space of E^3 (with some additional points attached), called nowadays elliptic and denoted S^3, in which the Vth postulate has been replaced by the following one.

V´´. Through a given point P not on a line l, there is no line in the plane of P and l which does not meet l.

So there are three different geometries for the space and investigations of Beltrami !1835-1900) and Felix Klein

(1849-1925) have made it clear that all three are logically equally valid. More precisely, if one is consistent, so
there are the other two. If we accept, e.g., the Euclidean
geometry as a basic one, there are models in it for both
the hyperbolic and elliptic geometry.

With the three geometries in mind, one can raise the
question: What is geometry (for the space)? The question
has been pursued by Klein in his well-known Erlangen Program from 1872 <3>. Klein related to each then known kind
of geometry the so-called main group of its basic transformations which leave fundamental geometric properties
intact. E.g., for the Euclidean geometry E^3 it is the
group of all rigid motions (rotations, translations, symmetries, and all their compositions). And Klein's answer
was: Geometry is the theory of invariants of the main
group. Since a group is an algebraic concept, one may
call this answer an algebraic one.

Another answer to the same question, what is a geometry?, may be inferred from the work of David Hilbert
(1862-1943) who showed that all three geometries (and some
others as well) can be put on a sound axiomatic basis <2>.
In that way geometry becomes an axiomatic theory developped on a basis of a set of axioms (somehow related to observed properties of the space or considered as "geometric"). Since such was the original structure of the Euclidean geometry in the "Elements", that answer may be called
classic.

The importance of distinction between geometries may
be exemplified by their implementation into modern cosmological theories. E.g., in Friedman's models for the Big-
Bang Theory the future of the Universe depends on the kind
of geometry of its spatial sections. According to that
theory, the Universe expands very rapidly at first, but
then the expansion slows down somewhat: in the case of

elliptic geometry the expansion actually reverses to become eventually a collapse, in the case of hyperbolic geometry it expands for ever, and in the case of Euclidean geometry the speed of its expansion slows down to zero <7>.

But one can raise another and apparently more important question, with the direct reference to Newton: What is Space itself? What it is mathematically? That question has been raised by Riemann in his famous qualifying lecture for the title of Privat Dozent in 1854 <8>. Riemann's ambition to make the lecture understandable for non-mathematicians resulted in its unprecise statements and total avoidance of mathematical symbols, and a consequence of it remains its mysteriosity. However, his another paper elucidated topic a little and efforts of his followers extended it to cover great areas of modern mathematics under the name of differential geometry.

Roughly speaking, Riemann's idea is twofold: firstly, one should define a substratum which underlies Space and consists of all its points together with some fundamental relations between them, more fundamental than metric ones (in the later development it was topology and topological relations that played that role, and substratum is what we call now a three-dimensional topological manifold); secondly, one should give the substratum a metric coming from the observation of the external world.

Manifold is a set which "looks locally like a Euclidean space". More precisely, n-dimensional manifold is a topological space (Hausdorff, with a countable base) each point of which has a neighbourhood U with the property that there exists a homeomorphic (i.e., one-to-one, surjective, continuous, and inversely continuous) mapping $f: U \longrightarrow R^n$.

Two-dimensional manifolds are called surfaces and have all been classified since long. E.g., closed surfaces are: sphere, torus (boundary of a doughnut), torus with two

holes, projective plane, Klein's bottle, etc. And a bit
more refined characterization proof leads to the remarkable discovery that each closed surface can be given one
(and only one) geometry: elliptic, Euclidean, or hyperbolic. Elliptic is sphere and projective plane, Euclidean -
torus and Klein's bottle, and all the other surfaces are
hyperbolic.

Three-dimensional manifolds have not been classified.
In fact, it is not even known whether a classification is
possible (for n-dimensional manifolds, where n = 4, 5, ...
it is not <5>). However, we know now a good deal about
them and there are ingenious tools to continue those investigations.

It is now rather commonly accepted that the substratum
of the Universe is, at any given time, a three-dimensional
manifold. Since we know the world only locally, metric put
on that manifold should have the form allowing local procedures and Riemann has proposed the form

$$ds^2 = \sum_i \sum_j g_{ij} \, dx^i \, dx^j \, ,$$

where ds is an infinitesimal increase of length and g_{ij}
are functions of a point defined by physics. Riemann's is
then the third approach to geometry and one may call it
a metric one.

From the Riemann's viewpoint, the absolute space of
Newton, i.e., the three-dimensional Euclidean space with
the Pythagorean metric, is the simplest kind of a manifold
with the simplest metric

$$ds^2 = (dx^1)^2 + (dx^2)^2 + (dx^3)^2,$$

i.e. $g_{ij} = 0$ for $i \neq j$ and $g_{ij} = 1$ for $i = j$. Any
other three-dimensional manifold can be viewed upon as
glued of some (possibly many) copies of E^3 and the global
metric may become in the process quite complicated.

Riemann's approach has turned extraordinarily succesful. A host of his followers (including members of the Italian geometric school like E. Beltrami (1835-1900), G. Ricci--Curbastro (1853-1925), T. Levi-Civita (1873-1941), L. Bianchi (1856-1928), but also E. Christoffel (1829-1900), O. Bonnet (1819-1892), F. Klein (1849-1925), S. Lie (1842--1899), and many others) have developped that approach to an important branch of modern mathematics. Hermann Minkowski (1864-1909), Albert Einstein (1879-1955) and Hermann Weyl (1885-1955) have connected it to the relativity theory and in this way differentla geometry became a mathematics of modern cosmology theories. Research in this area, one of the most vivid in contemporary mathematics, remains to be one of the most serious interest to mathematicians, physicists, cosmologists.

In the last decade, however, there appeared, rather unexpectedly, a new area of intriguely interesting research in geometry. It has started with the question: What kind of geometry can one put onto a three-dimensional manifold? Before proceed to answer that question one should first explain what does he mean by a geometry in a three--dimensional manifold. And William Thurston, who raised that question, does it by combining algebraic and metric approaches.

Since the substratum of the Universe (at any given time) is a three-dimensional manifold, let M be such a manifold. Physics tells us that the metric in the Universe is locally homogeneous (in each point the Universe looks the same) and complete (from each point in any direction one can send a light signal). So let our manifold M admit a geometric structure, i.e. let it possess a metric which is locally homogeneous and complete. Now turn to the topology which allows constructions called coverings, i.e. a construction, for a given M, of a three-dimensional manifold X with the projection mapping $X \longrightarrow M$ which is a local homeomorphism. Among all such coverings there is unique M^* called universal covering of M. Now any covering space of M inherits from M a metric such that the projection mapping is locally an isometry. Hence if M admits a geometric structure, so does its universal covering M^*. It is now a theorem of I. Singer <10> that a locally homogeneous metric on a simply connected manifold M^* must be globally homogeneous, i.e., the isometry group of M^* must act transitively. Thus we may regard M^* together with its isometry group as a geometry in the sense of Klein and in such a case we can sensibly say that M admits a geometric structure modelled on the geometry of M^*. Now Thurston's question can be specified in the following way: How many and what geometries may appear in the universal coverings M^* of three-dimensional manifolds M (with a metric locally homogeneous and complete)? The answer is: There are eight of them, namely three-dimensional Euclidean E^3, three--dimensional elliptic S^3, three-dimensional hyperbolic H^3, Cartesian products $S^2 \times E^1$ and $H^2 \times E^1$, universal covering $(SL_2 R)^*$ of the Lie group $SL_2 R$, Lie groups Sol and Nil (the latter is the so called Heisenberg group).

Of all eight geometries seven are rather well known. In particular, there have been classified all three-dimensional manifolds modelled upon one of them <9>. From the mathematical point of view, however, the most interesting and the least known is the eigth one which happens to be the three-dimensional hyperbolic space H^3.

It should be emphasized that not every three-dimensional manifold M has a geometric structure modelled upon some geometry (in the sense of Klein). E.g., no so-called connected sum, with the only exception of the sum of two copies of three-dimensional projective space, has a geometric structure in that sense. There is, however, an extremely interesting and powerful Geometric Conjecture of Thurston which says that each (closed, orientable) three-dimensional manifold can be cut into pieces (in a certain canonical way) each of which has a geometric structure modelled upon some geometry <11>.

Here we come to an end of the story which has traced, very briefly, one line of a development of modern mathematics with only roughly delineated interrelations and side developments. It seems, however, that in spite of all its shortages the told story reveals an interesting and often underestimated feature of mathematics. When Newton needed a space to build his mechanics, there was at hand the substratum underlying the Euclidean geometry (called since the three-dimensional Euclidean space). When Einstein needed a new kind of geometry to put his relativity theory to work, there was the Riemannian geometry ready to use. In these two cases (and one may quote many more) success was possible because mathematics was well ahead of all needs, evolving first, in a response to its own impulses, some sort of conceptual construction which turned to be afterwards mysteriously well fit to the requirements of science in her duty of describing the world. Mathematics is important because it is, to a high degree, autonomous.

References

<1> Boyer, C.B., The History of the Calculus and Its Conceptual Development, Dover Publications.

<2> Hilbert, D., Grundlagen der Geometrie, 7th ed., Leipzig 1930.

<3> Klein, F., Vergleichende Betrachtungen über neuere geometrische Forschungen, Erlangen 1872. Reprinted in The Mathematical Intelligencer 0 (1977), pp. 22-30.

<4> Kline, M., Mathematical Thought from Ancient to Modern Times, Oxford University Press, New York 1972.

<5> Markov, A.A., Unsolvability of the problem of homeomorphy, Proc. Intern. Cong. Math. 1958, pp. 300-306.

<6> Sir Isaac Newton's Mathematical Principles of Natural Philosophy and His System of the World, Translated into English by Andrew Motte in 1729, The translation revised, and supplied with an historical and explanatory appendix, by Florian Cajori, Two volumes, University of California Press, Berkeley and Los Angeles 1962.

<7> Penrose, R., The Geometry of the Universe, in the book: Mathematics Today, Twelve Informal Essays, Ed. by L.A. Steen, Springer Verlag 1978.

<8> Riemann, B., Über die Hypothesen, welche der Geometrie zu Grunde liegen, Abh. Ges. Göttingen, Math. Klasse 13 (1868), pp. 133-152.

<9> Scott, P., The geometries of 3-manifolds, The Bull. London Math. Soc. 15 (1983), pp. 401-487.

<10> Singer, I.I., Infinitesimally homogeneous spaces, Comm. Pure Appl. Math. 13 (1960), pp. 685-697.

<11> Thurston, W.P., Three dimensional manifolds, Kleinian groups and hyperbolic geometry, The Bull. London Math. Soc. 6 (1982), pp. 357-381. Reprinted in the book: The Mathematical Heritage of Henri Poincaré, Proc. Symp. Pure Math., vol. 39, Two parts, 1983.

Investigations of the Comet Halley's Motion:
Three Centuries in a Triumph of Newtonian Mechanics

K. ZIOŁKOWSKI

Space Research Centre, Warsaw, Poland

ABSTRACT

The discovery of comet Halley was a spectacular
confirmation not only of conjecture which revolu-
tionized the cometary astronomy but also of
Newton's theory of gravitation. Newtonian mechanics is a base of all investigations into the comet Halley's motion up to now. Short review of the
main works devoted to the modelling of the long-
term motion of this comet as well as discussion of
their results are presented.

Nearer to the gods no mortal may approach
E. Halley, "Ode to Newton"

1. HISTORICAL AND OBSERVATIONAL BACKGROUND

Among various celestial objects, comet Halley occupies
an unquestionable place in human history: it attracts and
fascinates the mankind for more than two millenia, from
three centuries it plays an unique and important role in
science and it has created an exciting opportunity for researches during last years.

Modern investigations of its motion began shortly after
the Isaac Newton's "Philosophiae Naturalis Principia Mathematica" has been issued. Edmond Halley, an admirer of
Newton, who not only edited this famous work but also co-
vered the cost of publishing it, employing Newton's laws

of motion and a method for computing parabolic orbits of
comets, determined the orbit of the bright comet of 1682.
In the letter of September 28, 1695, he told to Newton in
cenfidence that "... I am more and more confirmed that we
have seen that Comett now three times, since ye Yeare
1531 ...". Then, in "Astronomiae Cometicae Synopsis", publishing in 1705, Halley wrote: "Now many thing lead me
to believe that the comet of the year 1531, observed by
Apian, is the same as that which in the year 1607 was described by Kepler and Longomontanus, and which I saw and
observed myself at its return in 1682. All the elements
agree, except that there is an inequality in the times of
revolution: but this is not so great that it cannot be
attributed to physical causes. For example, the motion of
Saturn is so disturbed by the other planets, and especially by Jupiter, that its periodic time is uncertain to the
extent of several days. How much more liable to such perturbations is a comet which recedes to a distance nearly
four times greater than that to Saturn, and a slight increase in whose velocity could change its orbit from an
ellipse into a parabola? The identity of these comets is
confirmed by the fact that in 1456 a comet was seen,which
passed in a retrograde direction between the Earth and
the Sun, in nearly the same manner; and although it was
not observed astronomically, yet from its period and its
path I infer that it was the same comet as that of the
years 1531, 1607 and 1682. I may, therefore, with some
confidence predict its return in the year 1758. If this
prediction is fulfilled, there is no reason to doubt that
other comets will return." And in "Tabulae Astronometri-
cae", published in 1749, Halley pointed out once more "...
an agreement of all elements in these three, which would
be next to a miracle if they were three different comets;
or, if it was not the approach of the same comet towards

the Sun and Earth in three different revolutions, in an ellipsis around them. Wherefore, if accordingly to what we have already said, it should return again in the year 1758, candid posterity will not refuse to acknowledge that this was first discovered by an Englishman."

The comet was recovered on Christmas night 1758, 16 years after Halley's death, by Saxon amateur astronomer Georg Palitzsch from Prohlis near Dresden. Thus, Halley's hypothesis was fulfil and this comet is therefore named after him. The return of comet Halley as predicted became a source of vast scientific and philosophical implications for astronomy in general and comets in particular. It was also a spectacular triumph of Newtonian mechanics and one of the first natural confirmation of Newton's theory of gravitation.

Since 240 BC, the observations of comet Halley, made during the last 30 apparitions thereof, are documented in many historical records. Their usefulness in investigations of the comet's motion is very differentiated. While we can use in the orbit determination the astrometric positions from 1607 onward, all that we really know about the earlier apparitions are only the moments of perihelion passages as deduced from the analysis of European, Chinese and others chronical records (all the perihelion times of the observed apparitions are listed in Table 1). Those data constitute the observational base for the modelling of the motion of comet Halley. We can distinguish three stages in the investigation of this motion: old semianalytical methods for prediction of the past perihelion passages, numerical integrations of the comet's equations of motion which were based on the orbital elements linking last few apparitions of the comet, and a new approach enabling the use of the full observational material since 240 BC for

determination of the starting parameters for numerical integration.

Table 1

Moments of perihelion passages of the observed[1]
apparitions of Halley's comet deduced by Kiang
(from 1582 on, Gregorian calendar are used,
earlier dates are according to the Julian
calendar)

1986 February	9.46	837 February	28.27
1910 April	20.18	760 May	22.5
1835 November	16.44	684 September	28.5
1759 March	13.05	607 March	13.0
1682 September	15.27	530 September	26.7
1607 October	27.56	451 June	24.5
1531 August	25.8	374 February	17.4
1456 June	9.1	295 April	20.5
1378 November	9.02	218 May	17.5
1301 October	24.53	141 March	22.35
1222 September	30.8	66 January	26.5
1145 April	21.25	-11 October	5.5
1066 March	23.5	-86 August	2.5
989 September	9.0	-163 November	17
912 July	9.5	-239 March	30.5

2. METHODS OF COMPUTATIONS

The first attempts to recognize the past motion of comet Halley were connected with an identification of old
observations. Hind[2], by subtracting step by step the value
of mean orbital period of the comet from the known perihelion time and then adjusting the obtained moment to observations which were found around it, has determined the
all moments of perihelion passages from 11 BC. Angström[3],
using these data, has fitted the simple Fourier interpo-
lation formula to the orbital periods. His idea was further developed by Kamieński[4] who has derived from empirical evidence and theoretical considerations two independent formulae for the dates of perihelion passages and

the period between successive returns. They are taking into account main perturbations due to Jupiter and Saturn
in the comet's motion. The empirical evidence used as a
starting-point by Kamieński is the series of identifica-
tions by Cowell and Crommelin[5]. Using the variation of
elements technique with various limitations and approximations Cowell and Crommelin have obtained the perihelion
times back to 240 BC. Their computations were extended
back to 622 BC by Viliev[6]. It is worth to be noted that
the main purpose of his investigations was an attempt to
demonstrate that the biblical vision of the prophet Jere-
miah (1, 13-14) was an observation of Halley's comet in
623/622 BC.

In order to check the usefulness and reliability of
the above mentioned formulae in researches of the long-
term motion of comet Halley, Kamieński applied the "cyclic method" based on the commensurabilities in the motions
of the comet, Jupiter and Saturn. Although the orbital
period of the comet is variable and oscilates within the
limits of few years, it can be assumed that during the
long periods of many revolutions of the comet these in-
equalities are for the greater part smoothed out. The cy-
cles applied by him are intervals of time comprising in-
teger multiples of the periods of revolutions of the comet, Jupiter and Saturn about the Sun (e.g. 23 C = 149 J
= 60 S = 1768.1 years, where C, J and S denote the mean
orbital periods of the comet, Jupiter and Saturn respectively). They are used for the prediction of the comet
Halley's perihelion passages in a similar manner as does
the Saros in the prediction of eclipses.

Using the Fourier formulae and the cyclic method Ka-
mienski has computed the perihelion times of the comet
Halley back to 2312 BC. The results obtained by independent methods agree very well. Then he has extended yet

those investigations as far back as to the 10th millenium
BC[7]. While counting from the present days this encompasses
150 revolutions of the comet around the Sun. It is difficult to judge the reliability of these unique results in
the history of cometary investigations. Their originator
was fully aware of the fact that only observations can be
considered as an ultimate proof. Because of absence of
such proof Kamieński approached the problem in quite another way[8]. Namely, while assuming the genuinenes of the
moments of perihelion passages determined by him as well
as taking as probable that some descriptions of unusual
phenomenons on the sky as accompanying famous events from
the ancient history can be attributed to the apparition
of Halley's comet, he determined with relatively high ac-
curacy the dates of corresponding events. Since some of
them are determined by historians quite precisely then
their consistence with those being calculated for comet
Halley may to some extent justify the rightness of assumptions being made. In this manner Kamieński analyzed
thoroughly among others the King David's foundation of the
First Temple in Jerusalem (1010 BC), the fall of Troy
(1150 BC), the destruction of Sodom and Gomorrah (1757
BC), the birth of Abraham (1856 BC) and even the biblical
flood (3850 BC) and the catastrophe of legendary Atlantis
(9542 BC). Of course, the credulous acceptation of results
of such study as ascertained facts would be equally unjustified as definite rejection of them. They constitute an
interesting contribution to the chronology of ancient history and may be helpful in learning the comet Halley's motion in the very distant past.

All further works devoted to the problem of the motion
of comet Halley are based on the numerical integration of
the equations of comet's motion. In order to receive trust-
worthy results the equations must contain terms describing

not only the full planetary perturbations but also so called nongravitational effects, because the motion of comet
is influenced by more than the solar and planetary gravitational forces. Those effects decelerate the motion of
comet Halley about 4 days/revolution.

Taking into account an empirical term in the equations
of motion giving the effect of secular decreasing the solar attraction with time, Brady and Carpenter[9] have linked
by one system of orbital elements the four apparitions in
1682, 1759, 1835 and 1910. Although they made it by a
"trial and error" fit to the observations, the result was
used by Brady[10] as initial values to the numerical integration of the equations of the comet Halley's motion back
to 2647 BC.

Using a least squares differential improvement process
and applying the Marsden's model for nongravitational forces as a rocket-like effect acting on the cometary nucleus
due to the vaporization flux of water ice, Yeomans[11] has
found a set of orbits linking three or four apparitions
within the 1607-1910 interval. He has found that the orbit
based upon the 1607, 1682 and 1759 observations is the
best one for the starting to numerical integration. Then
Yeomans and Kiang[12] have integrated the comet Halley's
motion back to 1404 BC. In order to fit at the accurately
observed perihelion passage times in 837, 374 and 141,
they introduced some empirical adjustment to the computed
perihelion times and eccentricity in 837 when the comet
closely passed the Earth (to within 0.03 astronomical unit).

Similar approach to the problem we can find in the paper by Landgraf[13]. He has also improved the orbit of comet
Halley by the least squares method but using the positional observations from the six last apparitions over the
1607-1984 interval. A next important distinction is con-

nected with the model of nongravitational effects. Yeomans and Kiang used the nongravitational parameters as values being constant in time while Landgraf considered those as some linear function of time. After making some subjective change in the orbital elements also for 837 Landgraf extended the process of numerical integration back to 2317 BC.

A new procedure of orbit improvement elaborated by Sitarski[14], basing on the perihelion times as observational data to the least squares correction of orbital parameters, enables to use in a uniform way both modern positional observations as well as inaccurate ancient observations from which the perihelion times have been deduced. Thus the full observational material can be used to a linkage by one system of orbital elements the all apparitions of comet Halley since 240 BC. Using a secular change of the semi-major axis of the orbit as a model of nongravitational effects Sitarski[15] has found a parabolic time dependence in the nongravitational forces over the 87 BC - 1986 AD interval. Taking into account the 250 best positional observations from 1835-1986 period and the 26 observed perihelion times since 87 BC, he has obtained one system of dynamical parameters of the comet, which were used then as input data to the numerical integration back to 1457 BC.

3. COMPARISON OF RESULTS AND CONCLUSIONS

From among results of investigations presented here we selected for a comarison the most reliable ones which are representative for the discussed approaches to the problem of comet Halley's motion over long periods of time. As we can see from Table 2 the predicted by some authors past perihelion times outside the observational interval are different in general. It means that the true

Table 2

Moments of past perihelion passages of Halley's comet
predicted by some authors (in years BC)

Kamieński	Yeomans, Kiang	Landgraf	Sitarski
236.7	239.6	239.7	239.8
313.8	314.3	314.6	314.9
390.7	390.3	390.7	391.1
466.7	465.5	465.7	466.1
544.9	539.6	541.0	542.7
622.0	615.4	617.3	619.2
701.1	689.9	692.0	694.1
778.2	762.4	768.9	770.2
856.6	835.6	845.6	845.6
932.8	910.6	923.9	922.4
1009.9	985.1	1001.2	999.2
1085.3	1058.1	1081.0	1075.0
1162.3	1128.7	1158.5	1151.5
1238.2	1197.6	1236.7	1226.7
1316.4	1265.3	1315.7	1302.6
1393.3	1333.4	1393.3	1379.6
1472.3	1403.2	1472.8	1457.5
1549.6		1550.4	
1627.8		1628.0	
1704.1		1705.4	
1781.0		1782.2	
1856.2		1858.2	
1932.6		1935.2	
2008.3		2010.0	
2083.7		2086.8	
2160.3		2162.2	
2235.6		2239.0	
2312.9		2316.1	

motion of comet Halley in the distant past remains unknown.
However, the data summarized in Table 2 lead us to some
interesting conclusions.

A cause of discrepances between results obtained by
Yeomans and Kiang and those calculated by Landgraf or by
Sitarski, lies in the secular changes in nongravitational
forces[16]. It appears that this small factor plays an very

important role in the process of orbit determination and
should be carefully investigated. This surprising conclusion confirms also the fact that when Landgraf has dras-
tically reduced the value of a parameter characterizing
the linear variability with time of nongravitational effects[17], he obtained result being almost identical with
that obtained by Yeomans and Kiang.

An amazing agreement of Kamieński's and Landgraf's data seems to be especially interesting. It shows that there
is such a model of motion along with such initial conditions which, in the process of numerical integration, confirms earlier calculations made in simple manner and by
primitive methods. Is that only a fortuity? Today it is
yet impossible to answer this question. It is worth, however, to be stressed that the procedure used by Kamieński
has such advantage that in the long time intervals gives
some kind of "smoothening" of various small inaccuracies
and absences of a model of the comet's and planetary motions, i.e. levels the effects of those factors which in
the process of numerical integration of the equations of
motion are playing a particulary important role because
systematically increasing their values. It seems that in
the light of the last works and new possibilities of making full use of observational data being available as
well as better cognizance of the nature of nongravitational effects, this amazing agreement can bring us to the
recognition of the true motion of comet Halley and can indicate a direction of researches in the future.

REFERENCES

1. Kiang, T., "The past orbit of Halley's comet", Mem. R.
   astr. Soc. 76, 27-66 (1972); see also ref. 12; the perihelion time in -163 is given between November 9 and
   26 (Stephenson,F.R., Yau, K.K.C., Hunger, "Records of
   Halley's comet on Babylonian tablets", Nature 314, 587-

592), we took the middle moment.

2. Hind, J.R., "On the past history of the Comet of Halley", Mon. Not. R. astr. Soc. 10, 51-58 (1850).

3. Angström, A.J., "Sur deux inégalités d'une grandeur remarquable dans les apparitions de la comète de Halley", Nova Acta Reg. Soc. Sc. Upsal, Ser. 3, 1-10 (1862).

4. Kamieński, M., "Researches on the Periodicity of Halley's Comet", Part I - "Determination of the average Period of its Revolution", Bull. Acad. Pol. Sci. Ser. A, 101-140 (1949), Part II - "The Past of Halley's Comet (Preliminary results)", Bull. Acad. Pol. Sci. Ser. A, 33-38 (1951), Part III - "Revised List of Ancient Perihelion Passages of the Comet", Acta Astron. 7, 111-118 (1957).

5. Cowell, P.H. and Crommelin, A.C.D., "The perturbations of Halley's Comet in the past", Mon. Not. R. astr.Soc. 68, 111-125, 173-179, 375-379, 510-514, 665-670 (1907 -8).

6. Viliev, M., "Issledovaniia po teorii dvizheniia komety Galleia", Izv. Russ. Obshch. Lub. Miroviad. 6, 215 -219 (1917).

7. Kamieński, M., "Orientational Chronological Table of Modern and Ancient Perihelion Passages of Halley's Comet 1910 AD - 9541 BC", Acta Astron. 11, 223-229 (1961).

8. Kamieński, M., "Perihelion passages of Halley's Comet in ancient times, deduced from chronicler's notes and the author's theory of the comet", Bull. de la Soc. des Amis des Sc. et des Lettres de Poznań, Ser B, Livr. 18, 117-135 (1965).

9. Brady, J.L., and Carpenter E., "The orbit of Halley's Comet and the apparition of 1986", Astron. J. 76, 728 -739 (1971).

10. Brady, J.L., "Halley's Comet: AD 1986 to 2647 BC", J. Brit. astr. Ass. 92, 209-215 (1982).

11. Yeomans, D.K., "Comet Halley - the orbital motion", Astron. J. 82, 435-440 (1977).

12. Yeomans, D.K.,and Kiang, T., "The long-term motion of comet Halley", Mon. Not. R. astr. Soc. 197, 633-646 (1981).

13. Landgraf, W., "On the Motion of Comet Halley", ESTEC EP/14.7/6184 Final Report (1984).

14. Sitarski, G., "Linkage of 53 Observed Perihelion Times

of the Periodic Comet Encke", Acta Astron. <u>37</u>, 99-113 (1987).

15. Sitarski, G., "The Effects of Nongravitational Forces on the Long-Term Orbital Evolution of Comets", in "Comet Halley 1986, World- Wide Investigations, Results and Interpretations", Ellis Horwood Ltd., Chichester England, in press.

16. Sitarski, G. and Ziołkowski, K,, "Investigations of the long-term motion of comet Halley: what is a cause of the discordance of results obtained by different authors?", Proc. 20th ESLAB Symp., ESA SP-250 Vol.III, 299-301 (1986).

17. Landgraf, W., "On the motion of comet Halley", Astron. Astrophys. <u>163</u>, 246-260 (1986).

ISAAC NEWTON'S PHILOSOPHY OF NATURE

Zygmunt Hajduk

Department of Philosophy, Catholic University of Lublin,
20-031 Lublin, Poland

ABSTRACT

The issue of Newton's philosophy of nature may be considered historically or systematically. The historic contribution to a discussion of this subject could be a presentation of the so-called
case study (Fallstudie), i.e. the relevant episode in the history of science viewed in the perspective of the definite theory
of rationality and based on source materials, Newton's in this
case. In this paper we shall abandon the historic approach for
the sake of the systematic one. On the ground of the accepted
terminological assignations we shall "place" Newton's heterogeneous - as it will turn out - philosophy of nature on the network of its other conceptions*. We shall also show the different
ways of its reception and its influence on the development of
science and the philosophy of nature.

1. THE PHILOSOPHY OF NATURE AND THE PHILOSOPHY OF SCIENCE

The problems of the philosophy of nature (PN) generated, in different periods of the history of philosophical thought, lesser or
greater interest. In the periods of increased interest in the philosophical knowledge of nature the objective and metaobjective controversies are strongly marked.

In many centres, especially Anglo-American PN is not considered
a standard philosophical discipline. It is suggested that the term denotes either an isolated historic event in the form of Goethe's,
Schelling's, Hegel's Naturphilosophie, or a scholarly doctrine, a continuation of the tradition based on the contents of Aristotle's

"Physics", or a new tradition derived from the writings of A.N. White-head. It is believed in these centres that natural sciences are a reli-able source of information about the physical world. The philosophy of science, on the other hand, offers a logical (philosphical) interpreta-tion of a research process and its results. At its foundations are the methods and the results of empirical sciences. Both groups of sciences decide about the considered understanding of nature. In this way it is stressed that there is no basis for typically philosophical investiga-tions of nature. It is, therefore, believed that PN as a philosophical discipline does not practically exist now. Its place has been taken by the philosophy of science or the philosophies of particular sciences, e.g. physics or biology. They deal with the essentials of these sci-ences or science in general.

Questioning the need for the existence of PN together with natu-ral sciences is the result of perceiving only its supplementary role in relation to scientific knowledge of nature. The competition of the two kinds of knowledge being a remainder of the emancipation of natu-ral sciences from philosophy seems to be overcome nowadays. Objective investigations, on the one hand, and epistemological analyses on the other, complement one another and are indispensable for the adequate understanding of nature[1]. The role of PN rooted in different philos-ophies of science and in philosophies in science (Philosophie inner-halb der Wissenschaft)[2] as reflections and discussions on the given science manifests itself in a few typical situations. It precedes in time the formation of new branches of scientific knowledge but also participates in this formation. It is also one of the (heuristic) sources of constructing the so-called basic theories made up of (qual-itative) ontological assumptions formulated in the language of the appropriate branch of mathematics and of the research tasks and prob-lems of the branch defined by these assumptions[3].

In another trend of studies, not restricted to classical philos-ophy the need for this kind of philosophical science is not negated, although no homogeneous and generally accepted programme of philos-ophical cosmology is suggested or realized. The problem concerns the

contemporized version of the possibilities of philosophical studies of
nature, different from those of natural sciences and also from these
known from metascientific considerations. We make no mention here of
the significance of these studies in view of different contemporary
philosophical trends (classical, tending towards phenomenology, Marxism,
positivism in the form of e.g. mechanicisim or energetism).

The historical and contemporary versions of PN seem to compose
a whole spectrum on whose one end we find the autonomic PN practised
within the framework of the definite philosophical systems and on the
other PN treated as a special case of the philosophy of science.

In the latter version of PN two, relatively independent kinds
of studies may be distinguished. The first one is in principle the
philosophy of natural sciences (PSM) and concerns the widely understood logical analysis of the scientific procedures and the language
of these sciences. It is of descriptive-normative character. First, it
describes and explains the actual scientific procedures, also their
historical aspect. The evaluative-normative-projective component of
PSM is then treated praxeologically. When treated pragmatically it ex-
hibits the actual process of practising these sciencs, when apragmati-
cally - the product of scientific investigation. If, in these considerations, the scientific procedure helps only to exemplify the methodological structures, we talk about the external philosophy of
science (PSE). If it is the object of investigation and justifies the
existence of such structures, we talk about the internal philosophy of
science (PSI). PSE is of normative character, PSI - of empirical, which
goes together with its dimension in time. PSI, therefore, takes into
account the developmental aspect of science, i.e. the characteristic
methods of modifying theories in consideration of the encountered empirical and theoretical anomalies. That is why the methodology of a
contemporary physicist will, on no account, be identical with the methodology of physics from, e.g. the period of Galileo.

The contemporary philosophy of science takes up also another
kind of investigation. The basic objects, processes and relationships
on the physical world create the reference of some advanced scientific
theories. A discussion of the philosophical implications of such theo-

ries, a modification of some theses concerning philosophical positions accomplished as a consequence of the obtained results is another interest of the philosophy of natural sciences (PSN). At its (PSN) foundations are the above mentioned results and not the procedures applied to obtain them. Its objective statements (e.g. about the nature of time, interactions, or the discussion of the mind-body problem) differ from the components of empirical theories, also objective, in the method of justification (are not supported in the strictly empirical manner, involving the inductive or hypothetical-deductive method) and investigation allowing to construct a more synthetic and uniform picture of the material world. The discussion of its problems involves both the results of the relevant empirical sciences and the philosophical components. These problems are sometimes included (E. McMullin[5]) in the ontology of science.

Because of the objectivity of the above statements there is a direct connection between PSN and PN and an indirect connection between PSM and PN. The latter is illustrated by such categories as determinism, causality, law, mechanicism, vitalism, teleology or teleonomy functioning in some of the contemporary types of PN. They are based on the analysis of the metascientific characteristics of physics and biology. These connections are not of the identity type. PN cannot, therefore, be reduced to the scope of the philosophy of science not only when it concerns the formal aspect of scientific propositions (its analytical element), but also its contents (its synthetic element). Such reduction would impoverish the philosophy of nature. The problems considered by the philosophy of nature go beyond the frontiers of contemporary science. But their tentative solutions are still dependent on the actual advancement of science, so they bear the hallmark of the hypothetico-fallible character of science and its results. PN requires a sufficiently general point of view which makes it possible to consider the specific kinds of experience and to emphasize the role on nature. Its conceptualizations are historically conditioned and do not exclusively belong to the realm of natural sciences[6].

In the typology of the philosophy of nature the source of knowledge about nature and the justification of its theses must be taken
into account. In the first order philosophy of nature (PN_1) the source
is independent of the constructions of natural sciences and in the
justification of its theses we do not refer to the results of these
sciences, so it is called the direct justification. The second order
philosophy of nature (PN_2) is genetically and justifyingly based on
the current scientific theories, so the justification is indirect. In
the philosophy of nature of the "mixed" type (PNM) there are two kinds
of evidence: besides the results of natural sciences the theses of the
definite ontological and epistemological positions are explicite involved. In the construction of a more or less adequate picture of the
world definite scientific theories and philosophical view-points are
used.

The above differentiation of PN is based, first of all, on its
relations to natural sciences[7]. Besided the openness of PN to the results of natural sciences, philosophical, also logico-methodological
openness of natural sciences is also very important. The above differentiation seems to be accurate for at least one reason. If PN_2 is
considered uncontroversial, especially by positivism and Marxism, PN_1
and sometimes PNM are decidedly debatable. The problem of the existence
and character of the philosophy of nature is also focused on PN_1. Those
who consider themselves philosophers of nature practise PN_1 or PNM.
Practising PN_2 is treated as creating a certain version of the philosophy of science.

Referring the above types of PN to natural sciences also brings
up to date the problem of the valid source of insight into nature which
would be independent of the sciences. This source is the pretheoretical
knowledge of the natural world, given in original experience, characterized in this context according to the writings of E. Husserl, M.
Merleau-Ponty and J. Piaget. The qualification of the tasks and methods differing PNM and PN_2 from these sciences is also very important
(e.g. the demarcation criteria based on intuition, the degree of gen-
erality, the methods of justification). It is so, because one cannot

exclude the possibility of realizing a specific continuum containing
the elements of empirical generalizations, advanced theories of nature
and philosophies of definite categories, e.g. of time.

The above presented typology of PN can be referred to some historical systems continued in the 20th century. Among the model examples
of the philosophy of nature are the relevant fragments of the systems
of Aristotle, Descartes, Kant, Schelling, Hegel and their later conti-
nuations in scholasticism and new scholasticism, and also in the
writings of E. Husserl, H. Bergson, A. Eddington and K. F. von Weizsäcker.

Besides the above mentioned systems there also exist evolutional
philosophies of nature which take into account the time factor in explaining nature (genetic explanation). They exist in the above mentioned three forms and explain not only the evolution of the structures of nature in time. The idea of evolution functions as the basic
principle explaining their statements (e.g. the systems of H. Spencer,
S. Alexander, P. Teilhard de Chardin).

2. NEWTON´S POSITION

In the above conceptual framework Newton's PN is not a homogeneous system. The times of Galileo and Newton brought a change in the
understanding of the idea of nature. Nature is no longer the cosmos in
the sense of objectivistic PN of the Greek period. Nature as cosmos
realized mathematical proportions, regularity, unity, was a delibera-
tely well-ordered whole of which man was an integral part. In modern
times the Greek **physis** was replaced by the predicate physical. Nature
is a physical world treated as an idealized space-time continuum which
contains matter and which is represented mathematically. This world
is given in experience, experimentally. The notions of experience and
experiment functioning in natural sciences, on the ground of Newton's
empiricism may also be applied to philosophy. Modern PN is of empirical
character. The term "experimental philosophy" was used by Newton in
formulating the fourth rule of philosophizing. In the empiricism re-

presented by Locke, Hume and also by Newton the idea of the relationship between philosophy and empirical knowledge is shaped. According
to Newton there is no reason why the systematic knowledge of nature
and philosophy should be separated. The whole knowledge of this kind
is one natural philosophy. Its components which are not principles are
not epistemologically certain. Unlike Aristotle or Descartes Newton
does not consider it necessary to create an ideal of scientific knowledge in which obvious, primary principles would be its starting point.
His position, however, is not entirely different, for his mechanics
contains the element of notional necessity. This component was later
too much stressed by his advocates. Newton's characterization of
science is, at the same time, different from the similar attempts of
Boyle and Huygens who stressed its temporariness, probabilistic character and the lack of certainty. It would be risky, however, to distinguish Newton's philosophy of nature because of its peremptoriness
characteristic of, e.g. the system of Descartes. Newton frequently
stressed the tentative and verifiable character of science justified,
first of all, experimentally. If we, then, talk about Newton's philosophy of nature, it is not because of its a priori certainty or the
specific methodology which enables to create this philosophy, but because of its very general categories and the explanation of natural
phenomena making a coherent whole. The distinction between PN and science of which PN is a hypothetical generalization or a categorial conceptual framework is not clear in the empiricist tradition. Although
the questions concerning the scientific method can be abstracted from
it, they do not belong to the philosophy of nature but to the philosophy of science (PSM). Newton did not embark on the studies of this
kind, but like Galileo, participated in the disccussion on these problems resulting from the controversies caused by his theories in dis-
pute (the theory of gravity, the theory of light)[8].

Besides the metascientific component (PSM) Newton's philosophy
contains PN_2 and PNM. PN_2 consists of the philosophical consequences
of scientific theories. These consequences are not deductive but methodological; their justification is based on these theories. Accord-

ing to the positivistic orientation (E. Mach and his followers) it is
the only valid version of Newton's PN. Its propositions are logically
derived from the results of the studies being a mathematical analysis
of the results of observation. Newton's PN_2 can also be seen in the
perspective of its concepts of absolute space and absolute time. It
conditions the consistency of the principles of mechanics explaining
the phenomenon of motion[9]. Its justification is the predictive success of mechanics and the logical coherence of the construction of
The Principles which it inspired. Newton's PN can also be treated as
PNM when the origin, structure and evaluation of natural theories are
also conditioned by the statements of the definite philosophical position, especially by its ontology. The philosophical vision of the
world implicitly conditions a scientist's expectations as to those
determinants of the theories. The relationship between scientific,
philosophical and theological issues is illustrated by Newton's (S.
Clark's) controversy with Leibnitz concerning absolute space being
sensorium Dei[10]. Rejecting the Cartesian a priori method of constructing physics Newton is, at the same time, its follower in the
sense that he effected Descartes' purpose and constructed the deductive system of dynamics which was Descartes' idea and at which Descartes aimed[11].

3. THE RECEPTION OF NEWTON´S POSITION

The physicists of the 18th century (e.g. O. Goldsmith, W. s'Gravesande, H. Pemberton) attached importance not only to the cosmological
synthesis accomplished by Newton, but also to his new idea of science.
The latter did not influence the representatives of the classical British empiricism (Locke, Hume, Berkeley). K. R. Popper[12] points to
Berkeley's criticism of the Newtonian empiricism and inductivisim.
A change on that score took place in the 19th century. The epistemological implications of Newton's physics, his methodological ideas were
considered by nearly all representatives of British philosophy (among
others by J. Herschel, W. Whewell, J. St. Mill, W. S. Jevons, A. de
Morgan). It was Th. Reid who introduced the Newtonian ideas of causa-

lity, induction, hypothesis into the main trend of British philosophy.
He unconditionally accepted Newton's methodology, his rules of philosophizing demonstrating them in the context of empirical epistemology.
He pronounced himself in favour of inductivism and against the method
of hypothesis. The four rules of philosophizing, which are not accurately defined and justified[13] by Newton, characterize induction which
is realized in three steps: (1) making observations and experiments;
(2) formulating general laws on the basis of facts; (3) educing further statements about facts from these laws. This formulation of inductive procedures is based on measurements, it is then quantitative.
It is also different from Bacon's formulation of induction in that respect that their results are not considered certain, moreover, they do
not involve the method of predication (**praedicabilia: genus - species**)[14].

The first of the above rules[15] also concern the real, and not
imaginary (figments of mind) causes, such as the Cartesian whirls in
place of gravity explaining the movements of planets. The first rule
also points to the significance of the principle of simplicity which
functions in the framework of the Newtonian empiricism together with
the principle of nature uniformity. According to this position the components of science are provisional. Even the principles of natural
theories are not interpreted as the ultimate formulations of the
causes of these phenomena. The second of these rules is interpreted as
one of the formulas of the principle of nature uniformity which in this
case is the basis for analogical inference. The third rule is a version
of the two previous rules and serves other purposes[16].

Another interpretation of these rules is given by Whewell who,
contrary to Reid, does not accept Newton's methodology unconditionally.
He assimilates it to his own system of the philosophy of science in
which there are categories (e.g. necessary truth, induction as understood by Whewell: consilience of induction) that do not fit in the
framework of empiricism. Reid interprets the Newtonian idea of cause
from the empiricist's position. It is an observable antecedent event
which serves to formulate a law in the set of elementary conditions.
Whewell does not place causes among empirical events. Causes are con-

structions of a hypothetical system and serve to explain adequately the empirical data. Gravity is such a construction of the system, acceptable because of the different sets of data which are explained in this system. Whewell's interpretation of Newton's first rule is in accordance with the above remarks. The category of cause is replaced by acceptable explanatory theories; the metaphysics of causes should be separated from the logic of scientific systems. The second rule is considered redundant by Whewell. The next two rules (III and IV) concern the status of induction. According to Newton induction served to formulate laws on the basis of phenomena. Whewell emphasized the role of a certain idea in this process according to which these phenomena are expressed.

Reid and Whewell also disagree about the interpretation of Newton's fourth rule[17]. Reid points to the fallibility of inductive inference, to the possibility of their falsification based on the new results of scientific research. According to Whewell these inferences as laws are conclusive and the new results of scientific research influence the changes of the positions of the laws in the theoretical system by making them more precise and changing them, so that they allow for exceptional situations. In this connection the source of modifying the laws is not the same for Newton, Reid and Whewell. For Newton and Reid it lies in the new results of experiments, for Whewell in the attempts to coordinate them with the theoretical system on the ground of conceptual analyses.

Whewell's interpretation of the fourth rule is referred to by P. K. Feyerabend. It is characteristic of empiricism accompanying the development of science in the post-Galilean period. In traditional empiricism, or in theoretical monism characteristic not only of the narrowly understood inductivism (e.g. J. St. Mill), but also of the orthodox philosophy of science (among others R. Carnap, C. G. Hempel, E. Nagel) philosophers abandoned the theoretical aspect of science, especially when evaluating its results. The essence of the theoretical development was sought in formulating more and more general theories possessing a common set of empirical data. Feyerabend, like Whewell, does not consider himself a representative of empiricism understood in this way. He questions both the thesis of a common set of empirical data having no theoretical basis and the thesis that theories are tested exclusively on the basis of facts. Instead, he maintains that no theory can be questioned if there is no other competitive theory. Progress is a result of the competition between the new and the already existing systems. Theoretical monism must, therefore, be replaced by theoretical pluralism. The evaluation of theories is not empirical in the sense that it is based on the language of observation which is common to different theories. Extraempirical arguments are also involved in it.

Feyerabend also emphasizes the significance of this rule in Newton's methodology. First, it sanctions the two methods of modifying general statements or theories inductively inferred from phenomena. On the one hand they are made more precise which enriches their informative content, on the other hand their application is limited when confronted with exceptions such as the phenomena not taken into account when constructing a theory. Moreover, this rule suggests the asymmetry between the results of experiments and theories on the one hand, and the hypotheses on the other. General statements, also called theories, are formulated and confirmed on the basis of facts. Only then hypotheses are put forward, which serve to explain them, and on which the condition of the agreement with facts or theories is imposed. The experimentally formulated laws cannot be rejected only on the grounds that they are not in agreement with some hypothesis explaining phenomena[18].

The problem of the status of hypotheses is also controversial in Newton's philosophy of science. The term "hypothesis" is not univocal[19] and is applied both to observational and theoretical statements, and to statements of the fiction type. It denotes neither phenomena, nor the statements inductively inferred from them. It usually expresses an explanatory conjecture accepted without any experimental endorsement. Newton used his hypotheses consciously, e.g. in the context of the corpuscular nature of light. He did not accept

the position of R. Cotes, his follower and the author of the Preface to
the second edition of **Principia** (1713) who put forward a programme of
physics without any hypotheses. Newton does not exclude them from the
realm of science, they are its integral part. The next editions of
Optics contain the so-called queries and the edition of 1717 contains
31 of these queries. The hypotheses which they contain did not undergo
experimental testing, they should, however, be subject to investigation.
They are misapplied when they serve as the foundations for the acceptance or rejection of scientific theories. These foundations are created
by experimental evidence and not by the indirect exertions to justify
these hypotheses by eliminating the well-known alternative ones, as we
are unable to find whether all the alternative hypotheses were taken
into account. Only in this case the elimination of all hypotheses, except one, would testify in favour of the unfalsified one. The discrepancy between the results of experimental research and the accepted
hypotheses does not suffice to question these results[20].

The discussion about the hypotheses was also stimulated by the
well-known maxim: hypotheses non fingo (I frame no hypotheses) which
appeared in **scholium generale** of the third volume in the second edition
of **Principia**. It was discussed by historians of science (among others
by R. Hall, N. R. Hanson, A. Koyré, E. Strong, R. S. Westfall and
especially B. Cohen). Its positivistic interpretation (E. Mach), according to which Newton was not interested in the hypotheses explaining
the causes of phenomena but only in examining facts is relatively well-
-known. It must be noted that the maxim appears in the context of the
casual explanation of gravity and is not directed against the hypotheses concerning the real causes, but against the Cartesian fictions for
which there is no place in experimental philosophy. It is also a retort
to the criticism according to which the physical theory included in
Principia is not adequate because it does not offer the causal explanation of gravity. The point is, however, that the inductively detected
gravity really exists, functions according to a formulated law and
suffices to explain the movements of the heavenly bodies and earthly
objects. Universal gravity, therefore, exists factually and not hypo-

thetically, i.e. fictitiously. It is difficult to say, however, to what
extent this principle can be applied. In its extreme interpretation it
says that thanks to it the category of certainty eliminating the hypothetical component was introduced into physics. The content of the
fourth rule of philosophizing testifies against this interpretation.
The statements of physics are not unquestionable truths, sufficiently
justified. Instead, they are provisional approximations susceptible to
revisions and corrections made on the basis of further results of research. Like Huygens, Newton did not share the opinion that in the
understanding of nature science, following the example of formal sciences, could achieve the status of a reliable and definite discipline.
The role of hypotheses is not limited to discovering new directions of
theoretical and empirical studies. They also serve to explain the ascertained facts which is determined by certain limitations. They should
not be transformed into the assumptions of the system. Instead, the
speculations on the hypothetical causes of phenomena should be avoided.
Another interpretation of the maxim in question is connected with this
limitation. Its task was to prevent metaphysical speculations or concealed qualities (qualitates occultae). However, no support for this
limitation can be found in the text of **scholium generale**. Another interpretation of Newton's maxim recommends accumulating facts explained
by defining the relationships between them which are expressed in the
language of mathematics[21].

NOTES

[*] This conceptual framework is based on E. McMullin's, "Philosophies
of Nature", New Scholasticism 43, 29-74 (1968) and J.J. Compton's,
"Reinventing the Philosophy of Nature", The Review of Metaphysics 33,
3-28 (1979)

[1] This idea is expressed by A. Einstein who states that epistemology
and science are so related that without science epistemology would
be an empty schleme, and science without epistemology - if it is at

all possible - primitive and obscure. I. Lakatos' paraphrase of
Kant's maxim is analogous: the philosophy of science without the history of science is empty; the history of science without the philosophy of science is blind. Cf. A. Einstein, Bemerkungen zu den in die-
sem Bande vereinigten Arbeiten, in: Albert Einstein als Philosoph und
Naturforscher, Stuttgart 1955, 507; I. Lakatos, History of Science
and Its Rational Reconstructions, in: Boston Studies in the Philosophy of Science 8, 91 (1971). Also cf. B. Kanitscheider, Einleitung
und Übersicht, in: Moderne Naturphilosophie, Würzburg 1984, 9-10.

[2] Philosophy in science is propagated in Poland in the writings of M.
Heller, "How is Philosophy in Science Possible?" Studia Philosophiae
Christianae 22, 7-19 (1986), (in Polish); id., "Does the Real Philosophy of Nature Exist?", Studia Philosophiae Christianae 23, 5-29
(1987), (in Polish).

[3] The contrasting programmes of Newton (the corpuscular theory of
light, interaction at a distance, the thesis of vacuum: gravitational interaction between bodies takes place in the empty Euclidean
space) and Huygens (the undulatory theory of light, the theory of
local interaction, the theory of field) may be given as examples of
such theories compared to paradigms or to research programmes. From
the period of Newton and Huygens till the first quarter of the 20th
century the development of physics was dominated by the competition
of these programmes. They resulted in classical mechanics and electrodynamics. On the ground of SRT the competition between these programmes was partly neutralized. After modifying the branch of mechanics concerning local interaction it was possible to coordinate it
with classical electrodynamics. The incompatibility of the classical
theory of gravity and SRT created the need for a new research programme, similar in character to that of Huygens which resulted in
the birth of Einstein's theory of gravity (GRT). Cf. H. Törnebohm,
"Die Rolle der Naturphilosophie in der physikalischen Forschung",
in: Moderne. 20-22, 26-29, 36-37.

[4] It is also called the epistemology of science.

[5] "The History and Philosophy of Science: A Taxonomy", in: Minnesota Studies in the Philosophy of Science, Minneapolis 5, 24ff (1970).

[6] H. Lenk, "Homo Faber - Demiurg der Natur", in: Moderne. 112-114; S. Moser, "Der Begriff der Natur in aristotelischer und moderner Sicht", Philosophia Naturalis 6, 261-287 (1961); K. Hübner, "Wissenschaftliche und nichtwissenschaftliche Naturforschung", Philosophia Naturalis 18, 67-86 (1980); L. Schäfer, "Wandlungen des Naturbegriffs", in: Das Naturbild des Menschen, München 1982, 11-44.

[7] The philosophy of nature does not replace natural sciences, instead, it gives a more comprehensive analysis of their theoretical implications. Its programme consists of the following tasks: (1) the analysis of the accepted scientific knowledge, the reconstruction of the language in which it is expressed and, in particular, the formal reconstruction of scientific theories. (2) The generalization of these results in order to construct the model relationships between these disciplines and the categories best expressing the current development of science. (3) The contextual analysis of the relationship between theory and observation in order to discover the historical schemes of the development of science and the methods of explaining and testing. Cf. J. L. Esposito, "Reichenbach's Philosophy of Nature", Studies in History and Philosophy of Science 10, 189-191 (1979).

[8] J. J. Compton, Reinventing. 26-27; E. McMullin, Philosophies. 40-41, 47.

[9] E. McMullin, Philosophies. 45-46; M. Heller, Selected Issues and Trends in the Philosophy of Nature, Kraków 1986, 49 (in Polish).

[10] This fragment of the discussion is presented in the history of science by A. Koyré. B. Cohen, F. E. L. Priestley. The understanding of this controversy in the methodological perspective conditions, to some extend, the nonempirical factors affecting the choice of theories. The discussion of these factors based on the relevant suggestions of H. Hertz included in the extensive **Introduction** to The **Principles of Mechanics** may be found in the papers of G. Buchdahl and K. F. Schaffner published in the fifth volume of Minnesota Studies. 1970. Also

cf. R. Giere, "History and Philosophy of Science: Intimate Relationship or Marriage of Convenience", British Journal for the Philosophy
of Science 24, 285ff (1973).

[11] Having accepted the heuristic influence of philosophical ideas on
the theories of natural sciences we may find that the latter do not
realize these philosophical ideas in spite of their authors' inten-
tions. Further analyses of classical mechanics revealed that the
idea of absolute space does not function within its scope. Cf. M.
Heller, Selected. 109; E. McMullin, Philosophies. 45-48; R. S. Ingarden, "Descartes and Modern Physics", Kwartalnik Filozoficzny 19,
94 (1950), (in Polish).

[12] "A Note on Berkeley as a Precursor of Mach", British Journal for the
Philosophy of Science 4, 26-36 (1953). For Newton's inductivism op-
posed by Popper cf. A. Wellmer, Methodologie als Erkenntnistheorie,
Frankfurt/M 1967, 78.

[13] E. A. Burt, The Metaphysical Foundations of Modern Science, New York
1929, 215.

[14] L. Laudan, "Thomas Reid and the Newtonian Turn of British Methodological Thought", in: The Methodological Heritage of Newton, Oxford
1970, 102-131.

[15] R I: We are not do admit other causes of natural things than such as
both are true, and suffice for explaining their phenomena.

R II: Natural effects of the same kind are to be referred to the
same causes, as far as can be done.

R III: The qualities of bodies which can not be increased or dimini-
shed in intensity, and which belong to all bodies in which we
can institute experiments, are to be held for qualities of all
bodies whatever.

Quoted after R. E. Butts, "Whewell on Newton's Rules of Philosophizing", in: The Methodological. 134.

[16] Theories of Scientific Method, Seattle 1966. 130-139, 142-143,
E. Madden (ed.).

[17] Newton modified this rule many times (A. Koyré, Newtonian Studies,
London 1966, 269). Its ultimate version was included in **Principia:**

R IV: In experimental philosophy we are to look upon propositions inferred by general induction from phenomena as accurate, or very
nearly true notwithstanding any contrary hypothesis that may be
imagined till such time as other phenomena occur by which they
either be made more accurate, or liable to exceptions.

Quoted after P. K. Feyerabend,"Classical Empiricism", in: The Methodological. 166.

[18] R. E. Butts, Whewell on. 133-134, 146-148; P. K. Feyerabend, Clas-
sical. 159-160; A. C. Crombie, Medieval Science and the Origins of
Modern Science, Warsaw 1960, II, 391, translated from English by S.
Łypacewicz (in Polish). For further interpretations of Newton's
fourth rule cf. G. Buchdahl, "Gravity and Intelligibility: Newton to
Kant", in: The Methodological. 76.

[19] The semantic changes of this term are pointed to, among others, by
B. Cohen, Franklin and Newton, Philadelphia 1956.

[20] N. R. Hanson, "Hypotheses fingo", in: The Methodological. 14, 30;
A. C. Crombie, Science. 396; R. S. Ingarden, Descartes. 130; Theories. 124-128.

[21] N. R. Hanson, Hypotheses. 32; Theories. 119-123, 127-129; R. S. Ingarden, Descartes. 125, 127; A. C. Crombie. Science. 392-393. The
development of scientific knowledge is realized horizontally (growth
in surface) and vertically (growth in depth). The first one is called
Baconian and is accomplished through accumulating empirical evidence,
its description, generalization, systematization and prediction. The
second one is called Newtonian and is characterized by new ideas and
explanatory hypotheses going beyond empirical evidence and referring
to the unobservables. The epistemological depth goes together with
the ontological depth and is best realized when accompanied by its
logical organization of data. The proper development of knowledge
stipulates both types of growth. Growth merely in surface will be
hampered by the absence of explanatory hypotheses, while growth in
depth only could introduce into science an undesirable element of
uncontrolled speculation. Cf. M. Bunge, "The Maturation of Science",
in: Problems in the Philosophy of Science, Amsterdam 1968, 120-121.

NEWTON'S FIELDS OF STUDY AND METHODOLOGICAL PRINCIPIA

JAN SUCH

Institute of Philosophy
Adam Mickiewicz Uniwersity
61-674 Poznań, ul. Szamarzewskiego 91
Poland

ABSTRACT

Newton's research falls into three distinct domains: strictly scientific studies in mechanics, normal scientific investigation in the area of classical physical disciplines and parascientific (or pseudoscientific) studies in alchemy and theosophy. Accordingly, he made use of three different types of methodology. He was also aware of basic dissimilarities between scientific and parascientific analyses. However, he seems to have failed to recognize adequately the character of the methods which he applied in his research work in the first two domains.

1. SCIENTIFIC STUDY AND SELF-REFLECTION

A scientist need not be preoccupied with an analysis of his own study nor with general methodological questions. A self-reflective character of knowledge, i.e. its ability to study its own nature and methods by which it is acquired arriving, thereby, at self-cognition has always been a domain of the philosophical approach. Thus, since the time of Descartes, it has been customary to assess various branches of knowledge for their underlying philosophy taking into account their capability for self-reflection.

Scientists would hardly ever devote time and energy to pursue an analytical study of the methods which they apply and of the research activities which they undertake. In normal scientific practice, any reflection of that type would be more of a hindrance than real assistance |1|.

The above holds true, first of all, of the sciences. In the humanities and social sciences, self-reflection is far commoner; in such fields as sociology, political economy and humanistic psychology it is almost a routine procedure, which seems to be in agreement with the general belief of affiliation of these disciplines to philosophy.

There are times, however, when also in natural science self-reflection is common, when it becomes an obligatory component of scientific study. It happens so in the periods of crisis and historical turns or, in other words, when scientific revolutions take place. Then, outstanding scientists, who are aware of a critical situation in their discipline and whose responsibility it is to advance a revolution, become involved in philosophical disputes of a meta-scientific character - on ontology, epistemology, and even on axiology. This seems to have been the case of the physicists - authors of the twentieth century revolution in physics, i.e. cf Planck, Einstein, Bohr, Heisenberg, Schrödinger, Born, Dirac and Prigogine, who all have also been active philosophers.

Likewise in the seventeenth century - the age of the scientific revolution, when six great authors of modern science, i.e. Copernicus, Kepler, Galileo, Descartes, Newton and Leibnitz made a considerable effort to explain their procedures and methods of investigation. Two of them in particular, i.e. Galileo and Descartes, could correctly identify the methods which they were putting in practice. It seems paradoxical that the greatest of them all, Newton, who, telling the truth, has had no equals in the world of of science failed to understand the methods with the help of which he laid the foundations of modern physics, and which he partly invented by himself and partly borrowed from his contemporaries.

That gave rise to many biting remarks on the part of other physicists and philosophers. Einstein, for instance, claimed that Newton had been unable to see that the methods which he had been using with great success had nothing in common with the typical inductive generalization, contrary to the latter's belief that they were applications of the inductive methodology. Engels calling Newton "an inductive ass" might be hinting at the same |2|.

It is not easy to answer the question of why Newton failed to identify adequately the methods which he was using. Einstein was of the opinion that only in the twentieth century could physicists become aware of the methods of theoretical physics and describe them adequately. He believed that due to "the inductive illusion" that could not have occurred earlier. This may be true, but only to a certain extent. Newton's contemporary, Galileo, did, in fact, correctly define his own method as "geometric", thus pinpointing its most essential property.

Newton's was, undoubtedly, much worse situation than that of Galileo's owing to the flat empiricism which had been prevailing in his native England since the Middle Ages, having been strongly supported by such authorities as F. Bacon and R. Bacon. Therefore, it was more difficult for him to rid himself of "the inductive illusion" than for his continental colleagues |3|.

It seems, however, that there was still another reason why Newton was unable to overcome the illusion and identify correctly his methods. Namely, unlike Galileo and some others, Newton was active in several areas of study some of which extended far beyond mathematical physics. Distinct

methods which had to be applied in each of them by no means
facilitated the process of understanding their specific properties. As has been aptly remarked by Weizsäcker, science
is easier to be done than to be understood.

Furthermore, being involved in so many basically distinct activities, Newton must have been limited both in his
time and energy to be as much concerned with the methodological questions as some of his contemporaries (e.g. Descartes and Galileo).

II. THREE FIELDS OF NEWTON´S STUDY

Newton was actively engaged in at least three areas of
human knowledge. Only two of them may be said to belong to
science proper, the third one was merely in some "contiguity"
to it.

1. Studies in "the Classical Physical Sciences"

The first field of Newton´s interest comprised "the
classical physical sciences" which dated back to the Helle-
nic epoch. At Newton´s times they were geometry, astronomy,
geometric optics and mechanics. In T. S. Kuhn´s opinion,
the former three were the only branches of physics whose
language and methodology, founded already in antiquity,
made them inaccessible to laymen and, consequently, the
works written in each of those fields were comprehensible
to specialists alone |4|. According to Kuhn´s criteria,
they were well-developed scientific disciplines which,
having abandoned the questions of fundamental character,
could concentrate on puzzle-solving, entering, thus, the
"paradigmatic" (i.e. mature) phase of their development,
which is always marked by the true scientific advancement
consisting in presenting solutions to given problems |5|.

At Newton´s times the above mentioned group of the sciences of classical physics represented the entire (or nearly so) strictly scientific knowledge. Accordingly, we may say that the first domain of Newton´s scholarship belonged to **the sciences** or was **strictly scientific**. His major work <u>Philosophiae Naturalis Principia Mathematica</u> (henceforth, <u>Principia</u>) belongs in this area.

2. "Baconian" Studies

Next to the traditional branches of physics, i.e. classical physical sciences, the seventeenth century faced the growth of the so called "Baconian sciences" whose scientific status was largely due, in Kuhn´s opinion, to the emphasis which the philosophers of nature laid on experimenting and on compiling various natural histories, including a natural history of crafts |6|. The Baconian "histories" were mainly concerned with thermal, electrical, mechanical and chemical phenomena.

Although in the seventeenth century each of the two types of study was confined to a distinct geographical territory (i.e. England was the centre of the Baconian studies while the Continent, mainly France, was housing the representatives of the traditional disciplines), Newton was exceptional in making contribution to both trends.

The scientific revolution of the seventeenth century, in which Newton played the major role, radically transformed the classical physical sciences and led to the acceptance of the Baconian sciences. The latter had grown not from a scholarly university tradition but from the applied arts and craftsmanship. They were much dependent on the new experimental programme as well as on new instruments whose introduction was advanced by the craft.

At Newton´s times, the major difference between the two discussed "schools" lay in that the classical physical stu-

dies were applying mathematical methods and were based on
measurement and instrumental observation. Thus, they were
of a strictly scientific, naturalist character or, putting
it differently, they comprised the whole of natural science.
Whereas the Baconían studies were using experimental methods.

However, unlike the experimental methods used in ancient and medieval times, Bacon's methodology, whose most
eminent advocates were Boyle, Gilbert and Hooke, was not intended to confirm and exemplify the existing theories, but
rather to study nature in the conditions which had not been
investigated before or, even, in conditions which could not
normally occur.

While experiment was highly valued in the Baconian study, theoretical considerations were intentionally disregarded.
The scholars were clearly unaware of the mutual relationship
between the two. Hence, the Baconian "school" of the seventeenth century was outside the natural sc iences based on
mathematics.

Until the mid 1800's the Baconian tradition remained
undeveloped in the sense that its proponents could not propose a coherent theory which would have any predictive power. According to Kuhn, its evolution as well as the character of its publications reveal a striking resemblance to
many social sciences of today, yet they are totally unlike
the classical physics of their contemporaries |7|.

At the turn of the eighteenth and nineteenth centuries,
the Baconian studies acquired (at least in physics and chemistry) the scientific status thanks to works of such
authors as Aepinus, Cavendish, Coulomb, Gauss, Poisson, La-
voisier and a few others. But only during the second scientific revolution, which came at the beginning of the nineteenth century, could Baconian physical disciplines undergo
a transformation similar to the one which had affected classical physics some two hundred years before, during the first

revolution. The quantitative approach became obligatory both
in experiments (measurement) and in theoretical   studies
(quantitative theories), which is best evidenced in works
of such scientists as Fourier, Clausius, Kelvin and Maxwell.

Strangely enough, throughout all that time,   including
the nineteenth century, the two "schools" of the   physical
sciences, i.e. classical and Baconian, were developing independently of each other. A few exceptions such as Newton
and some other scholars who contributed to both traditions
could not alter the overall picture. Generally   speaking,
classical disciplines formed "mathematics" (in fact, they
were quasi-mathematical disciplines) whereas Baconian studies were treated more like "experimental philosophy".

Not only the founder of the Baconian school, but also
many of his followers such as Franklin, Black and   Nollet
had never placed much trust in mathematics nor in the quasi-
-deductive structure of classical physics; therefore   they
never even attempted to acquire any mathematical skills.
That was certainly an important subjective reason for keeping them apart from the classical discipline.

The reverse relation was far more complex. Owing   to
the newly introduced experimental equipment, the Baconian
movement exerted some influence upon the older, classical
disciplines, in particular on astronomy. The results, however, was rather a gradual increase in experimental rigour
on the part of the proponents of classical physics than any
substantial change in their orientation. As most adherents
of the Baconian tradition did not confide in mathematics,
so, many classical physicists were often reluctant to appre-
ciate the significance of empirical experiment and, accordingly, they used it rather sparingly. Needless to say, Newton was an exception, together with some continental scientists (e.g. Huygens and Mariotte) who like him belonged to
both traditions.

Another factor which contributed to the separation of the two schools was the fact that, unlike the Baconians, classical scientists were by no means amaterus. They were scholars with academic status, most of them university employees and members of the Academy |8|.

The long lasting split within the physical disciplines into those of the classical and those of the Baconian traditions was additionally strengthened by the independence of the mechanical arts from the non-mechanical ones. The latter became a subject of the scientific interest much later than the former, and had far more utilitarian objectives.

In spite of the above mentioned numerous factors whose importance cannot be underestimated, the integrating tendencies were already quite strong in the nineteenth century. As a result, the twentieth century physics has emerged as a well-developed unified science. Two processes seem to have had a crucial role here, namely, mathematization of Baconian disciplines, on the one hand, which transformed them into strictly scientific research, and on the other, the introduction of experimental methods into the classical physical sciences.

3. Parascientific Studies

Equally active was Newton in the third area of this studies which not only exceeded the boundaries of natural science but, in fact, lay outside any science. We shall call them activities of a parascientific character. They ranged from alchemy through the attempts to explain gravitation and the nature of absolute space, to the hypothesis of "the first push", and the role of God and of supernatural phenomena in the physical world.

Typical of this tendency was his treaty De gravitations et aequipondio fluidorum (hereafter, De gravitatione) written in the 1660´s. He discussed in it, for the first time, the conception of immaterial (unsubstantial) ether, which could be recuirring in a number of his later works. In the treaty, Newton departed from the idea of material ether, but not for good; in other numerous works and letters he would be again considering the possibility of explaining certain physical phenomena in terms of ether, either material or immaterial, i.e. spiritual. In <u>De gravitatione</u> Newton attributed the function of material ether to God Himself, initializing, thereby, his search for the divine medium or, in other words, for the "unsubstantial ether".

Surprisingly enough, in the same work he introduced his concept of absolute space and, concurrently, the idea that the absolute space was the physical attribute of the infinite presence of God ("immanent effect of God", "constituted by the infinite presence of God"). Thus, next to rational justification of the absolute space hypothesis which came from Newton´s belief that the laws of motion would be void unless they were referred to some specific (absolute) system (the justification was first noted by Mach in his critical analysis of Newton´s famous pail experiment), Newton also presented the theological justification of the hypothesis.

In another work, <u>De aere et aether</u> (written in the 1670´s), he discussed the hypothesis of "action at a distance".

There is no doubt that in his parascientific studies Newton was under a strong influence of neo-Platonic metaphysics and More´s mysticism.

III. ON THE INTERRELATIONSHIP BETWEEN NEWTON´S METHODOLOGY
AND DIFFERENT AREAS OF HIS STUDY

The fact that Newton was engaged in three distinct a-
reas of study leads to the conclusion that his methods must
also have teen of various types, regardless of the degree
of his awareness of how much those areas were dissimilar |9|.
There must be differences in the approach to problems of a
strictly scientific character (Newton´s first field of interest), to questions of "normal" scholarship (his second
area of activity) and, lastly, to para- (or even pseudo-)
scientific studies of the third type which hardly ever com-
ply with any scholarly standards being closer to mystical,
theological or theosophical hypotheses. Thus, it seems only
plausible to claim that depending on the field, he had to
apply three distinct types of methods, i.e. strictly scientific, scholarly and para-(or pseudo-)scientific.

Newton´s unpublished works provide enough evidence that
he was fully aware of at least one of those distinctions,
namely, that between general scientific methodology and para-
-(pseudo-)scientific methods. It is not difficult to distinguish between those of his writings which are of a "phenomenalistic" character and the remaining ones which deal
with "philosophical" questions of the nature of the investigated reality. The latter are often claimed to belong to
works pursuing the ether hypothesis. Newton´s aims and approach to the subject-matter are clearly dependent on the
character of study and differ considerably from one type to
the other.

The phenomenalistic works (e.g. Principia or Optics)
which continue the scientific traditions (in both of the a-
bove senses) are marked by the mathematical rigour. They
contain very few unjustified hypotheses and heuristic pre-

misses, in accordance with the principle Hypotheses non
fingo (with emphasis on the last word). Instead, they are
full of detailed descriptions of various experiments and
their results. They are not intended to provide ultimate ex-
planations of, for instance, the nature of gravitation or
of absolute space; neither do they introduce base hypotheses.

Unlike the former, the works of the "philosophical" type
(such as De gravitatione and De aere et aether) are devoid
of any mathematical discussions and calculations, neither do
they contain any analyses of experiments. They introduce,
in that plase, many hypotheses and suppositions as to the
final causes and tend to present the ultimate interpretation
of the facts of nature, mainly of the theological character.
Newton's conceptions of both "material" and "immaterial"
ether are deeply embedded in his metaphysical speculations.

It is worth noticing that the above mentioned fundamental differences between the two types of work are best documented in his writings on absolute space. While the former studies deal with the purely physical problems, analy-
sing the functions which the absolute space would perform
in the physical world, the latter concentrate on the theosophical questions attempting to determine the "divine" nature of the absolute space.

For Newton, who could not forsake the idea of searching
for the ultimate theological interpretation of natural phenomena because he was deeply convinced that the utmost rationality of nature comes from God's rational project for
the world, it must have been extremely difficult to remain
confined in his "phenomenalistic" works to problems of the
strictly scientific character.

For that reason, the discipline which he was able to
impose upon himself in his scientific studies is one more
evidence of his greatness, greatness not only of his intellectual genius but also of his personality.

N O T E S

1 W. Heisenberg recalled once his and Weizsäcker´s failed
attempt to approach a herd of goats with a movie-camera.
Niels Bohr made the following comment in that connection:
"The goats managed to escape only because they were una-
ble to ponder and discuss how to do it." |see Heisenberg,
W., **Część i całość**, PWN Warszawa, 179 (1987)|. Although
in the goats´case it was rather instinctive behaviour
than scholarly activity, the analogy seems quite suitable; self-reflection, in particular if it is pursued
while a given action is just being carried out, is very
likely to reduce its effectiveness.

2 See Engels, F., **Dialektyka przyrody**, KiW Warszawa, 211
(1953).

3 Engels, criticizing the one-sided inductive approach,
writes "All that inductive cheat |comes| from the English
|...|". See Engels, **Dialektyka...**, 236.

4 See Kuhn, T. S., "Tradycje matematyczne a tradycje eks-
perymentalne w rozwoju nauk fizycznych" |Mathematical
versus experimental traditions in the development of
physical science| (in:) **Dwa bieguny. Tradycja i nowator-
stwo w badaniach naukowych** |The Essential Tension|, PIW
Warszawa, 73 (1985)

5 See Kuhn "Tradycje matematyczne...", 73.

6 See Kuhn "Funkcje pomiaru w nowożytnej fizyce" |The func-
tion of measurement in modern physical science| (in:) **Dwa
bieguny...**, 301-302. In Kuhn´s view, these sciences are a
new byproduct of the Baconian principles of "the new philosophy".

7 See Kuhn "Tradycje matematyczne..." 86.

8 See Kuhn "Tradcje matematyczne..." 92-93.

9 It seems possible to think of a higher number of those
areas. One could question, for instance, grouping together (i.e. as "parascientific studies") such dissimilar
activities as alchemy and theosophical speculations.
While the former, in spite of being often inspired by
metaphysics, were basically of experimental character,
the latter had little in common with experimental study.

METHODOLOGICAL ASPEKCTS OF NEWTONIAN REVOLUTION

Wacław MEJBAUM
Stettin University, Department of Humanities,
POLAND

ABSTRACT

I shall consider the Newtonian revolution as an issue of
contemporary science. The methodological basis of presented analysis appears Poincare,s philosophy and its conti-
nuations, have been elaborated in the 20-century "non-statement view". Some comments to Popperian hypothetism are
added.

1. The modern epistemology reveals the complete impo-
tence in understanding of classical physics. Let us consider the Kantist dualism of synthetic and analytical propositions in science. There is no satisfactionary answer to
the simple question: is the statement 'f = ma', synthetic,
or analytical ? Does this formula give an example of arbitrary definition /the definition of force or the definition of mass, what you like better/, or rather an example
of empirical law ? The discussion of the given problem has
been long and exciting one, but in vain - if you omit some
interesting remarks in papers of Moritz Schlick.

The very problem can be formulated as follows: what
are we to cenceive by the phrase "Hypotheses non fingo" ?
As I can see, there are two alternative interpretations.
The first one appears historically valid, but methodologi-
cally empty, the second one will be the main matter of

further considerations. In a historical sense there are no doubts that Newton wanted to exclude hypotheses recalling hidden matters or forces, convealed for experimental control. This idea, although rational, appeared rather difficult to explicate. The difference between "hidden" and "evident" matters is too indistinct, indeed.

Therefore we shall pass to the second interpretation. It depends on the opposition hypotheses and principles. The concept of hypothesis is then understood in contemporary meaning of the word. Hence the problem rises: how the principles can be discovered, where to seek guarantee of their validity ? It is quite clear that Kantist conception of statements, "synthetic a priori" is not useful in the case of principles of mechanics. It is also clear that these principles cannot be interpreted as inductional generalisations of evidence. Such a diagnosis leads us to Poincare's philosophy.

2. Let us consider the Poincare analysis of "phosphorus". Let us imagine the situation of chemist, who finds a piece of substance, which melts in temperature 45°C. All other properties of this substance are identical with properties of phosphorus. There are two possible decisions: a/ the temperature of melting for phosphorus can be different from 44°C and b/ there exist two kinds of substance, the first one - design it by P_1 - melts in 44°C, the second - in 45°C, assumed other characteristics to be the same. It is remarkable that Poincare preferred the second decision. In this meaning, he refuted a kind of primitive empiricism, but the label of "conventionalist" is also unfitting to him. In reality Poincare's device depends on supposition that scientific development can be reduced to changes in conceptional basis of science. These changes, however, are forced by evidence. The very core of science consists of existential statements or presumptions.

128

In the face of these facts, Poincare's philosophy appears as the first design of contemporary non-statement view. This problem requires a peculiar consideration.

3. The non-statement view has bound with the name of
J.D. Sneed. The essence of this view can be reduced to
statement that every law of physics appears an "open formula". The scope of application of such a formula is relied on the mode of interpretation of terms. To fix the
given formula is valid we need the clear definition of
domain of its semantical interpretation.

Where the principles of mechanics are satisfied ? The
simple answer is: they are satisfied in an abstract domain
which is defined as some kind of point-space. Is the physical world identical with a space of this kind ? It is
another question which can be examined only by using of
experimental methods. The evidence has revealed - for
example - that four-dimensional Minkowski space offers the
better aproximation to reality than the Galileo space,
assumed by Newton. In the face of this fact the replacement of Newtonian kinemathics by STR became the necessary
step. Changes of this kind will be repeated in the future,
however the Newtonian method of constituting physical
theories remains up-to-date.

Martin Heidegger was the first great philosopher who -
yet in thirties - noticed that the main task of modern
science depends on producing the own domain of exploration
or - in Heideggers language - "exploitation". In this
sense physics has began from building the space of material points assuming some conditions of mathematical type,
e.g. the condition of twofold differentation of every
function to be applied in theory. The discovery of this
domain appears the real meaning of Newtonian revolution.

In considerations concerning on the problem of truth
and falsehood in science, we consequently ought to differ

two relations: the relation of satisfying, linking the scientific theory, and its methematical model /let's say: a "superreality" founded above the actual world/ and the relation of co-ordination, which arises between a construct in "superreality" and the given fragment od reality. The science discovers "abstract truths" with reference to mathematical "superreality" and "concrete ones" with reference to material world. The relation of "concrete truth", is to be meant as the set-theoretical product of relations of satisfying and of co-ordination.

Such a schema allows to understand standard difficulties that have been met with by a student who wants to solve an exercise in elementary physics. Let it be the following problem: try to compute the trajectory of a missile binding two arbitrary chosen points in the space, /assumed the conditions are given/. Let the student know the relevant part of theorethical mechanics. However, he will not start without a decision in certain simple matter: ought he to consider the Earth as a material point, or as material solid, or - perhaps - as an infinite attracting plane ? But such a decision must be based rather on some kind praktice, than on perfect acquiantance of textbooks. Problems of this kind arise always when the solution of task depends on finding a relevant co-ordination between abstract and concrete objects, or, - using the words of Hans Reichenbach - finding convenient "coordinative definitions".

4. The last problem which I am to deal with in this essay, is the problem of decomposition. Let assume a scientific theory based on the set of statements: P_1, P_2, ..., P_n. Let now the evidence show that the theory is untenable in certain domain of application. There is a bit of possibilities:

a/ to refute the whole of theory i.e. the conjunction

$$P_1 \wedge P_2 \wedge \ldots \wedge P_n$$

b/ to hold some statements and to refute the other ones. In this case further remain some alternative decisions.

As I can see, the main vice of Popperian hypothetism is the lack of understanding of the the problem of decomposition. In fact, whenever a theory fails in explanation of evidence, the choice of a part of theory being to be modificated is very troublesome. It is one of the cosequences of the very fact that Popper's division of statements into falsifiable hypotheses and arbitrary definitions cannot be longer defended. Instead of this division I propose to agree /following W.O. van Quine in this matter/ that the principles of physics refill some double roles: they firstly define an abstract domain of theory and - secondly - they point out a possible scope of its application. Assuming such a point of view the Newtonian maxime "hypotheses non fingo" gains its contemporary sense.

Therefore the brute incoherency of theory and evidence in no case is sufficient reason to refute the theory. But - to be sure - it can indicate that a modification is needed. Consider the following example. What is the content of the inertia principle ? On the abstract level this principle claims that in four-dimensional space there are straight lines or geodethics. It is true in wide class of spaces considered by physicists. In an application to reality it means that every free move is to be dealt with as a geodethic in the relevant space.

Here remains exactly one problem which requires an empirical solution. Can all free moves be represented as geodethics in the same abstract space, or - perhaps - the choice of relevant space depends on "nature" of the moving object ? In the second case the physics would be no longer developed in Newtonian style. But such a fall appears as equally uncreditable to-day as it was 300 years ago.

EPISTEMOLOGICAL OBSTACLES
IN THE SCIENTIFIC DEVELOPMENT

Zdzisław CACKOWSKI

Institute of Philosophy and Sociology,
M. Curie-Sklodowska Univ., 20-031 Lublin, POLAND

ABSTRACT

Science is not a sum of individual discoveries added to
earlier knowledge. Earlier knowledge is always both a
means (instrument) for acquiring new knowledge and an
obstacle, because every truly new knowledge is also a
new way of thinking, and the truly new way of thinking
involves the modification and often the rejection of the
existing modes of thought, and the process of overcoming the traditional methods cannot proceed unhindered.

Three centuries have now passed since the appearance of
Isaac Newton's great work. **Philosophiae Naturalis Principia
Mathematica** provoked a revolution in the history of science.
One man was the author of this work. But is it possible to
accomplish such a radical change single-handed? There is no
doubt that the work of many earlier and contemporary scientists had contributed to Newton's achievement. Copernicus,
Kepler, Galileo, Hooke, but also Bacon, Descartes, Locke and
Hobbes are only some of the best known contributors to the
intellectual climate of the seventeenth century which played
a part in the development of the new theory and without which
it could not have spread. And how many less known figures
there were, and how many of those who have been completely
forgotten?

In the last decades of our century the study of the development of science has more and more directly emphasized the significance of the socio-cultural conditions of its development. The socio-cultural environment is considered both as a creative and as a hindering factor in relation to the development of science; some of the components of environment function as a kind of inspiration, others as an obstacle. An individual act of discovery in science is beginning to be considered today as not altogether autonomous. This is what is meant when one speaks of "Newton" as the author of Newtonian mechanics. The inverted commas are used to restrict the individual character of the discovery. This ascertainment presents us with a new investigative task of some importance for the history of science; which factors in the seventeenth century intellectual, or even more generally, cultural climate favoured Newton´s achievement and which acted against it; what kind of resistance had to be overcome by the emerging discovery?

It is said that every great theory is born as a heresy and dies as a superstition. My modest knowledge of the history of science makes me convinced of the validity of the above view. In what phase of its life does Newtonian mechanics find itself today? Has it completely become a part of the history of science just like his optics has? How have the particular elements of this theory left the stage of living science, or are they still in the process of leaving it? It is known that classical mechanics offered strong resistance against relativist physics and so, at the moment of the emergence of the new theory, it acted as a superstition, an active superstition opposing the new "heretical" theory. Has this resistance stopped? Have the categories of absolute time, absolute space and absolute mass ceased to function as obstacles in the development and spread of relativist theory? Does the active presence of classical physics function today only as a factor of opposition in relation to the new physics? Have the fundamental categories of classical physics retained a creative function, too?

I think that at least for me it is easier to speak of the hindering function of classical physics in relation to the development of contemporary physics than to point to its positively creative functioning. I think that the positively creative period of this theory belongs to the past but that it has not yet become a part of the history of science; it continues to be an active partner in the development of contemporary physics as a factor of opposition, as an active epistemological obstacle.

I borrow the latter term from a French philosopher, Gaston Bachelard (died 1982). The term "obstacle epistemologique" (epistemological obstacle) denotes for Bachelard the active resistance offered by the previously acquired knowledge, and the mode of thinking associated with it, in relation to new knowledge and a new mode of thought. The thing is that the efficiency of any mode of thought, and any mode of action, the efficacy of any algorithm gives rise to a certain inertia of thought, gives rise to the impression that it is effective in a broader area than the original one where it has proved its effectiveness. Thus the cognitive efficacy of the knowledge that has been gained so far and the mode of thought that has been elaborated gives rise to the generalization of this mode of thought, which can be characterized as intellectual usurpation. This usurpation functions as a creative factor until it reaches the limits of its actual efficacy but beyond these limits becomes an obstacle. It functions as an obstacle when new modes of thought, new algorithms must be sought for new areas of facts incompatible with the existing way of thinking. Then the formerly efficient mode of thought functions actively as an obstacle of intellectual development, an obstacle in the development of science.

If this is so then one may suppose that Newton, too, must have overcome the resistance of both the intellectual culture he had inherited and the climate in which he made his discoveries. Indeed, it was so. Now, I will try to support this supposition with a few examples.

In those times - the seventeenth and eighteenth centuries - philosophers and scientists lived in a strange atmosphere pertaining to the methods of valuable cognition.

The tradition of Aristotelian physics is a tradition of common-sense (everyday) empiricism coupled with the "eternal" principles of reason which contradict it, i.e. the "absolutely obvious" axioms (e.g., in a moving system the downward movement of a body that has been thrown up must deviate from its original course in the direction opposite to that of the movement of the system, a part is always smaller than the whole of which it is a part). On the other hand, the heritage of theological-religious thought of the Middle Ages which did not describe but rather evaluated the world in moral and religious terms continued to exert considerable influence. The latter approach supported and confirmed Aristotle´s suggestion of the fundamental difference between the earthly, "mundane", sublunary world and the world above the moon, the world of extraterrestrial (heavenly) bodies. The latter worlds was supposed to be perfect and the former imperfect. Thus, the two worlds could not have been governed by the same laws. It is well known how great was the opposition that Kepler had to overcome before he finally succumbed to the descriptive data and accepted the elliptical character of planetary trajectories. The resistance came from the traditional belief that the planets - heavenly bodies - must follow the perfect trajectories, i.e. circular ones.

It may be supposed that Newton had to deal with analogous obstacles. Usually, when reflecting on the ways in which Newton had arrived at the formulation of his mechanics, we consider the premises which played the most important role in the discovery of the simple dependence between the attracting force and the product of the masses of the two bodies involved, and the inverse dependence of this force on the square of the distance between them. Let us consider the role of Kepler's laws and Hooke's ideas in connection with the above. Meanwhile, equally, if not more, important - because more difficult - was the generalization of Newton's propositions. He applied his propositions to all phenomena of a certain type; he rejected and overcame the traditional division of the world into two worlds, each governed by different laws: the laws of physics were, for him, universal, there was no fundamental difference between the terrestrial and superterrestrial worlds, the "lower" world and the "higher" world. For the second time in the history of thought the same transformation of the style of thinking was taking place. The first such transformation was effected by Ionian philosophers (the "naturalists") who broke with the "vertical" type of thinking which divided the world into two non-equivalent parts: the superior sphere (existing above), the source of the truly important LAWS and TRUTHS, and the subordinate, mundane, earthly, everyday sphere; all the elements of the whole world were treated as equal, equivalent, and subjected to the same laws, and generating these laws in the same way. In later times - especially in the Middle Ages - the image of the world in man's thought became again divided and vertical; it was divided into the world "below" (earthliness of little importance) and the world "above" (the only authentic source of TRUTHS and LAWS). Newton once again "reunited" the world theoretically. Thus, he had to overcome a great resistance of traditional ways of thinking and traditional culture.

This gives rise to the question as to whether this resistance had been overcome by Newton alone? We have suggested that the answer to this question seems to be negative, that the efforts of many other scientists and philosophers must be taken into account. But at this point another question arises: were there any extra-scientific premises at work, too? I believe that this was the case: after all, the vertical culture was at that time attacked from many different sides - it was undermined by social and economic changes (the decline of the feudal socio-economic structures with their clear-cut division into "lower" and "upper" classes; the emergence of new socio-economic relations of capitalism based on the principles of equality and partnership), socio-political changes (the transition from structures in which the source of political authority was "above" to structures in which authority came from the mundane, social source), finally, the changes in religious culture, which also questioned the idea of the heavenly origins of hierarchical power. All these changes depreciated the ABOVE as an absolutely different world and were conducive to thinking aimed at overcoming the absolute opposition between the above and the below, between the superterrestrial world. In this way "mountains" and "valleys" came closer together in the newly emerging style of thinking, they were uniting to form one world, they were coming together at the opposite poles, but these opposite poles were parts of the same world, governed by the same laws. Thus, the changes in the style of thinking of the entire culture of the times turned out to be in agreement with Newton's vision of the world subjected to uniform and universal laws.

The above remarks may seem strange to someone who tends to regard the development of science as the result of the individual efforts of a single scientist, or the effect of the efforts of many individual scientists, representatives

of a given branch of knowledge. In fact, the intellectual
culture of science is always a part of the larger culture
and for this reason the dynamics of science cannot but be
coupled with the dynamics of culture as a whole. Thus, the
development of science must be considered, on the one hand,
as the process of overcoming the resistance within science
itself (the resistance caused by the existing knowledge and
modes of scientific thought) and, on the other, one must
also inquire into the premises of the dynamics inherent in
extra-scientific culture, the premises of cultural obstacles as well as the premises of cultural dynamics, trans-
ferred into the area of scientific thought.

We should add that scientists themselves, including
great discoverers like Newton, are not always aware of the
character of their own achievement. I shall exemplify this
with Newton's own motto: "hypotheses non fingo" (I shall
not make hypotheses). I can well suppose that this motto
denoted his opposition to scholastic speculations. We know
now that he failed to realize this project, or at any rate,
he was not entirely successful, because his theory of absolute space bordered on theological speculations.

The matters look even worse when his motto is treated
as a project for constructing science in a radically empirical way because, as we know, the fundamental propositions
of Newton's mechanics, or at least some of them, are non-
-empirical. This can be clearly seen, for example, when we
take a closer look at the first principle of Newton's mechanics: "Every body continues in its state of rest, or of
uniform motion in a straight line, except so far as it may
be compelled by force to change this state". In order to
understand the non-empirical character of this proposition
one only has to compare it with two earlier analogous formulas. A similar idea appeared already in Democritus; a
body in vacuity moves perpetually because there is no suf-

ficient reason for it to stop at any given point   rather
than at another one, so it cannot stop at all. There is no
doubt that Democritus' thesis is speculative because it is
based on a speculative assumption that atoms exist in vacuity (absolute vacuity). And here is an evidently empirical formulation of an analogous thought given by Leonardo
da Vinci: a body set in motion retains this motion for a
certain period of time. Here we are dealing with an evident generalization of an empirical observation. And the
first principle of Newton's mechanics? It certainly is not
a generalization of an empirical observation because it refers to an ideal object; it does not refer to any empirical
object accessible to observation. Is it then just as speculative as Democritus' formulation? It is difficult for me
to answer this question but undoubtedly it is not an empirical thesis. At any rate one thing is obvious: Newton's
way of practising science does not correspond to his methodological declaration in which he renounces hypotheses not
only in the sense of constructing suppositions but also,
and most of all, in the sense of making propositions which
are not subject to observational verification.

    In conclusion, I would like to say that the factors
of resistance emerging from the older scientific and cultural traditions that Newton had to deal with were diverse
and often acted in opposite directions. Primitive empiricism of Aristotelian physics counteracted the emergence of
the theoretical tendency which - as we know now - was a
necessary premise of the emergence and development of modern science. Galileo had to struggle against this tradition and so did Newton. On the other hand, scholastic spe-
culativism constituted a major obstacle on the way to combining theoretical considerations with experience and experiment. Modern science needed speculation; highly abstract
considerations, but such as might find the way to empirical

and practical applications, and which might be of use in dealing with experimental data.

<u>Professor Stanisław Szpikowski´s questions</u>: (1) Does the claim that every theory emerges as a heresy and dies as a superstition apply to the philosophical-ideological doc- trine professed by the author of the paper? (2) The thesis according to which the development of science is connected with overcoming obstacles ("obstacle epistemologique") does not apply to science itself; there are no such obsta- cles in science. Indeed, they might have functioned when science was bound with theological or philosophical thought. At present, science has gained independence from these ex- trascientific forms of culture and as a result its develop- ment is no longer connected with overcoming such obstacles.

<u>Reply</u>: As to the first question, my answer is unhesitating- ly affirmative. This thesis has a general significance, it is universal, or so it seems to me. No ideology, or world- -view, neither the one that I profess, nor any other, in- cluding the one that the questioner holds, is an exception. Only the life-span of each of them can be different, and the forms of their decline may be different, too. It may happen that the content of a given world-view has long been dead but the form survives, either as an empty cultural form or as a vehicle for new contents which may have only a very superficial link with the former one. The longevity of doctrines of this kind is, in such cases, only illusory, a given form or institution (or institutions) may become per- manent when it is employed to realize different interests (not necessarily its own). Then a single form may contain different doctrines, each of which corresponds to the needs of different epochs.

Anyway, doctrines which are truly alive are not, and

cannot be, free from death. The absolute permanence of man-
-made products is always a lie. It resembles a theory which
aspires to explain everything, everywhere and always. Only
a theory that is fundamentally w r o n g (contradictory or
empty) may explain everything everywhere. If a given theory
"explains" everything, it is a sufficient proof that it explains nothing.

With regard to the second question I must say that professor Szpikowski is twice mistaken. I apologize for the
world "mistake" but if I were more courteous in my use of
words I would have had to subject my own knowledge and my
own convictions to courtesy which would have been highly
inappropriate in a scholarly discussion.

First of all, the suggestion that science is not involved in philosophical disputes is erroneous. It is enough
to look at the texts of Einstein, Bohr, Heisenberg, Weizsäcker, Jacob, Monod, or Lorenz to see how important this
involvement is, and to what extent theoretical considerations of science lead to the so-called philosophical problems. Philosophical aspects of these considerations are
not an addition to scientific investigations but result from
them, and are organically bound with them.

Secondly, the supposition that science, free from all
theological or philosophical involvements, does not produce
its own obstacles to its further development is also erroneous. It does produce them and cannot fail to produce
them. The thing is, that every human activity (and its efficacy) has its limits beyond which it becomes ineffective.
But as long as a given type of efficacy functions within
the limits of its effectiveness, its effectiveness seems
broader than it really is. Thus, when problems appear from
outside the area of the effectiveness of a given method of
activity (including the scientific methods) then the efficacy which has been attained earlier claims to be effective

in the other area, too. This, in effect, constitutes an obstacle to the emergence of a new method. This law has the same force as the law of gravitation.

I had an opportunity to deal with this problem on the basis of scientific data devoid of any philosophical associations, i.e., Kekulé´s discovery of the structural formula of benzene. It is a most instructive example. Kekulé had earlier been involved in solving many "analogous" problems. He had found effective structural formulas of hydrocarbon compounds. All his earlier solutions produced open structures which suggested that open structures are the universal structures of hydrocarbon compounds. And then he had to face the problem of benzene. The first attempts at the solution tended towards the open structures. The persistence of this thought pattern which so far had proved effective blocked an earlier solution. The mechanism I am referring to here is the universal mechanism of every human activity and thought. It pertains to every man - from a schoolboy to a Nobel prize winner.

Science is not a sum of individual discoveries added to earlier knowledge. Earlier knowledge is always both a means (instrument) for acquiring new knowledge and an obstacle, because every truly new knowledge is also a new way of thinking, and the truly new way of thinking involves the modification and often the rejection of the existing modes of thought, and the process of overcoming the traditional methods cannot proceed unhindered.

Psychologists claim that the better one solves a task of the type: "Arrange 3 matches into an aquilateral triangle with its sides equalling the length of a match, arrange 5 matches into 2 such triangles, 7 into 3, 9 into 4, 11 into 5, etc., the more difficult will he find the task of making 4 such triangles with 6 matches. In psychological terms it involves the same difficulty as the one faced by Kekulé.

The mechanism of this game is one of the mechanism of human thinking, the thinking of a child and the thinking of a scientist, including a physicist. In short, a physicist also speaks prose. Does a physicist need the knowledge of this fact? At any rate, it does not bring him any harm.

WHERE THERE SCIENTIFIC REVOLUTIONS IN PHYSICS BETWEEN NEWTON AND EINSTEIN?

WŁADYSŁAW KRAJEWSKI

University of Warsaw, Dept. of Philosophy,
POLAND

ABSTRACT

In contradistinction to the usually held view, I claim that in the XIX Century physics there were scientific revolutions connected with the overcoming of the mechanistic philosophy: admittance of non central forces, of physical fields, of statistical laws of Nature, of non visual theories. They led to changes in the scientific methodology.

The expression "scientifc revolution" is popular
since famous book by Thomas Kuhn /cf. 1/. He criticizes
the traditional cumulativism according to which the growth
of science is a cumulation of knowledge /addition of new
truths to the old ones/. He shows that scientific revolutions occur, which replace old theories by new ones. However, Kuhn goes too far in his anticumulativism, considering a revolution as a complete break of continuity,
treating theories divided by a revolution as "incommensu-
rable", etc. /cf. 2/. But the existence of revolutions is
indubitable.

In general, not only "cumulative changes" occur in
the history of science but also "anticumulative changes",
i.e. refutations of some statements previously admitted
by the scientific community. There are also other types
of anticumulative changes, e.g. disappearance of some concepts from the langauge of science /cf. 3/. Of course, not
every anticumulative change is a revolution. When new
measurements lead to a new value of a constant, hence to
the abandonment of the previously admitted value, nobody
will speak about a revolution. Analogously, when a discovery of a new chemical element or a new planet leads to
the abandonment of the previons view about the number of
elements or planets. Hence, the question arises: when we
should speak about revolutions?

Physicists and historians of science usually hold the following view. In the XVII Century there was an initial revolution which created the classical physics /Galileo, Newton/. In the XVIII and XIX Centuries the classical physics evolutionary developed. Only in our Century new scientific revolutions occured: the relativistic one /Einstein, Minkowski/ and the quantic one /Planck, Einstein, Bohr, Heisenberg/. Some physicists even say that relativity was the last stage of the classical physics and only quantum mechanics made a revolution. In other words, in the whole history of physics there were only two or three revolutions.

This view uses a concept of revolutions as a general reconstruction of science. I prefer a more liberal concept. We have to do with a scientific revolution when an essential thesis, previously admitted by the scientific community, is refuted, and when this refutation leads to a change in the scientific methodology, i.e. when a methodological norm is replaced by another one.

If we use this concept we must admit scientific revolutions in the XIX Century physics. Besides of "local" revolutions in particular branches of physics /optics, heat theory/, there were also general revolutions which affected the whole physics, changing its essential methodological norms; namely, the four following revolutions:

1. Discovery of non central forces /Oersted, Ampère/ - refutation of the dogma that all forces are central.

2. Admittance of physical fields /Faraday, Maxwell/ - refutation of the dogma that all physical objects are bodies with definite borders.

3. Discovery of statistical laws of Nature /Maxwell, Boltzmann/ - refutation of the dogma that all laws of Nature are strictly deterministic.

4. Creation of a non visual mathematical theory of electromagnetism /Maxwell/ - refutation of the dogma that all phenomena should be explained by means of visual mechanical models.

Each of these discoveries led to a change in the methodology of physics:

1. The norm that we must search for central forces as causes of acceleration or deformation was replaced by the norm that we must search for central or non central forces.

2. The norm that we must search for bodies as bearers of forces - by the norm that we must search for bodies or fields.

3. The norm that we must serach for deterministic laws of all phenomena - by the norm that we must search for deterministic or statistical laws.

4. The norm that we must create visual models of all

phenomena - by the norm that we must create visual or not visual /mathematical/ models.

Each revolutionary has his predecessors. In science, too, a new idea often arises long before the revolution in which it wins. Statistical considerations may be found in works of D. Bernoulli and others; however, only Maxwell and Boltzmann discovered statistical laws which were commonly admitted by the scientific community. Huygens and others developed the idea of field /aether/; however, only after Faraday this idea has been commonly admitted in physics. In general, we speak about a revolution not then when a new idea arises but then when it wins - in society and in science.

All four above mentioned revolutions were connected with the overcoming of the mechanistic philosophy in physics. They may be considered as four steps or four aspects of the "antimechanistic" revolution. However, their contents were different, they led to the abandonment of different norms; therefore, I prefer to speak about four revolutions in the XIX Century physics which had a common denominator - "antimechanism".

All these discoveries had a great philosophical impact, may be with the exception of the first one. The second of them led to serious changes in the concept of matter, the third one - in the concept of law /and determinism/, the fourth one - in the concept of scientific explanation.

After the creation of quantum mechanics new revolutions in physics occured. E.g., the discovery of nuclear forces /Yukawa/ was a revolution because it led to the refutation of the electromagnetic theory of matter /cf. 4/. I may add that it led to the abandonment of the norm that in microphysics we must always search for electromagnetic interactions.

REFERENCES

1. T.S. Kuhn, The Structure of Scientific Revolutions, Univ. of Chicago Press, Chicago 1962.

2. W. Krajewski, Correspondence Principle and Growth of Science, D. Reidel, Dordrecht 1977

3. E. Pietruska-Madej, Continuity and Anticumulative Changes in the Growth of Science, in: W. Krajewski /ed./, Polish Essays in the Philosophy of the Natural Sciences, D. Reidel, Dordrecht 1982.

4. K.R. Popper, The Rationality of Scientific Revolutions, in: I. Hacking /ed./, Scientific Revolutions, Oxford Univ. Press, Oxford 1981.

ON THE ORIGINS OF MODERN EXPERIMENTAL SCIENCE

Zenon E. Roskal (M.A.)
Catholic University of Lublin
The Faculty of Christian Philosophy
The Faculty of Natural Philosophy, POLAND

Significantly, historical and philosophical attempts to reconstruct the picture of modern experimental science in its static and dynamic aspect are unsatisfactory. Subsequent versions do not withstand criticism due to the unsatisfactory answers they provide. Also, the demarcation line that divides old and new physics seems to be very unclear because there is no demarcation principle which separates the old Aristotelian-Thomistic philosophy of nature from the new philosophy of nature represented by Galileo and Newton.

It is customary to point to the following constituent features of the modern experimental science: empiricism, mathematical method, anti-dogmatism, the last of which is closely connected with broadly-understood rationalism. Most frequently, the works of Copernicus (1543), Galileo (1591) and Newton (1687) are pointed out as demarcation points. On the other hand, some historians of science, e.g. Duhem, look for the demarcation line in the 13th century. Others go much earlier still.

This status quo calls for a characterization of mod-

ern experimental science which would give insight into
its essence in both static (synchronic) and dynamic
(diachronic) aspects. One feels that from its essential
features one can naturally derive the boundary line which
separates the old from the new. The aim of this article
is to propose such a characterization and the resulting
demarcation principle.

When we analyze the works of unquestionable founders
of new physics such as Kepler, Galileo or Newton, we see
that apart from many differences which obscure the general picture there is one intriguing convergence. Namely,
all of them criticized Aristotle according to Plato's
conception or, to put it more accurately, all of them
supported the idea of employing mathematics in the description of real world. Needless to say, this was in the
spirit of Plato and against the spirit of Aristotle. Another characteristic convergence stems from their univocal attitude towards the legacy of ancient times.
Kepler puts it simply:

"In theology one must observe the opinion of
authorities, in philosophy however one must
observe the value of evidence."

This viewpoint - which goes back to the Renaissance philosophers of nature such as Paracelsus, Telesio, Bruno -
was grounded by Pascal and Descartes and finally became
the motive force and the very essence of the new. One
could ask the question of the conditions of such an attitude to the investigation of nature. Generally speaking,
it must be pointed out that the Aristotelian physics was
passing through a deep crisis at the turn of the 16th
century.

The conception of the heliostatic world put forward by
Copernicus, undoubtedly aided by Neoplatonic inspiration,
was completely incompatible with Aristotelian physics.

The acceptance of Copernicus' conception implied the search for new physics and, what is more significant, new metaphysics. The new method and, simultaneously, the physics were found by Galileo and Newton, while a metaphysics turned out to be readily at hand. This was the Pythagorean-Platonic metaphysics which was brushed aside by the Aristotelian mathematics.

It is the amalgamation of such physics, whose metaphysical core, characterized by independence and freedom of action which forms the heart of change and, simultaneously the essence of modern experimental science.

Thus we can adopt the following thesis: Modern experimental science appears to be a reaction of Pythagorean- Platonic metaphysics cropping up in the atmosphere of turning away from the authorities. The far-reaching physical, philosophical and methodological implications of these have shaped the history and the status quo of today's physics.

Diachronically, this ammounts to saying that the development, or, more broadly, the history of natural science was closely related to learning a lesson from its basis, i.e. Pythagorean-Platonic metaphysical core. Lack of space does not allow us to develop this argument further, but the sketchy presentation given above seems to support our viewpoint.

The characterization of the synchronic aspect of natural science shows that those factors which seemed to be constitutive appear, in the light of the above thesis, to be secondary: The empiricism and anti-dogmatism of modern physics is a natural consequence of the rejection of authorities. In the situation when one cannot adduce the evidence of authorities the only criteria of argumentation which remain are reasoning and experience. In other words, having rejected the authorities which fetter it,

science wins back the latitude of development and wide
perspectives.

The problem of the mathematical method is very much
similar. If we adopt the Pythagorean-Platonic metaphysics, it is natural and necessary to employ mathematics.
If we assume that it is idea-number which organizes reality, then the only means of finding it is the consistent
application of mathematics in the investigation of nature.
If we investigate nature from the quantitative point of
view, we get to its fundamentals, in contrast to the
Aristotelian-Thomistic philosophy of nature, which enables us to get to some aspects of nature. More generally,
metaphysics is not restricted to an aspective expression
of reality, but it reaches back to its fundamentals,
providing at the same time adequate means of describing
nature.

Diachronically, the characterization of natural
science shows that these aspects which were in agreement
with the metaphysical core have been preserved and became well-rooted in contrast to those rejected elements
which were incompatible with the metaphysical core. The
best example is provided by the historical development of
the concept of mass.

All sensual factors associated with this concept have
been falsified and rejected. A totally formal approach
which consists in braking symmetry in Higgs' action,
which makes its appearance in modern physics, is wholly
compatible with Pythagorean-Platonic metaphysics.

Let us now tackle the question of the demarcation
principle. Notice that any attempt to answer such
questions involves a good deal of arbitrariness, but when
one decides to systematize chaos, one must put up with
arbitrariness.

It goes without saying that the demarcation principle

should not be sought before Copernicus because then nobody fostered such an attitude towards ancient times.
Both Buridan and Oresme were critical about Aristotle,
but what they aimed at was not to eliminate Aristotle,
but to adjust his views to remain truthful to the facts.
Likewise Copernicus belongs to this tradition. Even if
we take no account of the fact that Copernicus' way of
argumentation is similar to that of scholastics, one
notices that Copernicus goes by the principles of symmetry and harmony. Although he uses mathematics in the
description of the world, he does not reflect on the
physical causes of this harmony.

Kepler turns out to be the first philosopher, in the
new meaning of the word. His work entitled Astronomia
Nova. AITIOΛOΓHTOE. Seu PHYSICA COELESTIS, Tradita
commentariis de motibus stellae MARTIS appeared in 1609.
The very title shows that his astronomy aims to look for
causes.

There are many arguments which may be given at this
point. First of all, notice that Kepler uses mathematics
in the description of reality, but, at the same time, he
enumerates physical causes of the existing symmetries and
harmonies. His standpoint is very much against authority
and, more particularly, against Aristotle's arguments. In
fact, it is he who breaks the age-old tradition in the
name of overcoming the Aristotelian-Thomistic metaphysics:

> "My first error was to take the planet's path as
> a perfect circle, and this mistake robbed me of
> the more time, as it was taught on the authority
> of all philosophers and consistent in itself with
> metaphysics." 1

One gets the impression that formulating the law of
elliptical orbits, Kepler accepts Plato, but nevertheless
breaks a Platonic postulate. This is not true however. The
law of elliptical orbits should be perceived as the

realization of a thoroughly interpreted postulate of
Plato. The subsequent laws are the best proof of it,
because old, superficial symmetries were replaced by new
ones which go deeper and are of greater significance.
Kepler's astronomy and physics run close to the spirit
of Plato or, to put it more precisely, they are a natural
consequence of the Pythagorean-Platonic metaphysics which
Kepler adopted. Therefore, Kepler's work constitutes a
turning-point and, we assume, the demarcation line which
divides old and new physics.

To sum up, one can venture a characterization of
modern experimental science which enables us to grasp
its significant constituent elements. This characterization is based on the thesis that modern experimental
science was founded on and developed from the base of
the Pythagorean-Platonic metaphysics, verified by the
criteria of reason and experience. It follows from this
characterization that the date 1609, when Kepler's
work appeared, should be treated as a demarcation point.

[1]Quotation taken from A.R. Hall. The Scientific Revolution 1500-1800: The Formation of the Modern Scientific
Attitude. London:1954.

CAN SCIENCE FREE ITSELF FROM THE NEWTONIAN PARADIGM?

Alina Motycka
Institute of Philosophy and Sociology
Polish Academy of Sciences
Nowy Świat 72, 00-330 Warszawa, Poland

ABSTRACT

The title question will be examined with respect to the following problems: the concept of experience: the problem of the visualness of science; the description of the state of the system; the objectivity of scientific description; the indeterminism and causality of contemporary physics; certain consequences of a non-traditional approach to these problems.

INTRODUCTORY REMARKS

The term "Newtonian paradigm" as used in the title denotes classical physics along with its methodological equipment. On the other hand, non-classical physics (sometimes also called "post-revolutionary" or "new" physics as compared with classical physics) means physical theories diachronically arranged in relation to the Newtonian theory.

In this sense, we speak of a diachronic ordering of old
and new theory.

In the present controversies over the relation between old and new theory in the philosophy of science,
a dichotomy of standpoints can be observed. Whereas some
recognize the participation - understood in one way of
another - of old theory (the existing knowledge) in acquiring new knowledge, others deny any relationship between
the two. However, even a supporter of the first of these
standpoints can take a critical attitude toward the manner
and value of the argumentation proposed. This question is
dealt with in the last part of this paper.

The problem raised in the title is one of the fundamental problems of the philosophy of science in the 20th
century - a continuously disputable and topical problem.
For it concerns the question of how knowledge is developing. This fundamental epistemological question has to be
answered anew in the 20th century, for it is being raised
by science itself (that which occurred in science in the
precisely by science, and not e.g. by contesting social
groups dissatisfied with science, or by some cultural phenomena.

That is why specific source material has been utilized
in this paper? it is the philosophical reflection of the
father and coauthors of 20th-century physical sciences. Obviously, only selected fragments of this rich heritage will
be considered, chiefly those concerning "reports" from la-
boratories on the cognitive situation in which the authors
of 20th-century breakthroughs in physical sciences found
themselves.

The title question is divided throughout the paper
into several more specific questions, the ordering of which
is in principle unimportant. They have been ordered in such

a manner that in replies to later questions earlier considerations could be referred to.

Defining in a general way the crux of the question
under consideration, one can say that it is an attempt at
an epistemological insight into the cognitive situation
of contemporary (post-revolutionary) physics, a specific
situation in which the very act of cognizing an object
influences this object, or in a way coconstitutes it.
Therefore, a situation will be considered (which has
arisen in new science) in which the real object undergoes
a change in the course of being cognized, and as a result
of this cognition it acquires certain properties. These
acquired properties are subjectively, i.e. externally,
conditioned.[1] In other words, in addition to the properties that this object possesses "in itself", it acquires
certain properties external with respect to itself, that
is, subjective properties. It is therefore in this sense
that one can generally speak of the subjective element in
cognition, that is, an element relativized with respect
to the cognizing man.

Considering that such is the cognitive situation in
post-revolutionary physics in the 20th century, two
additional explanatory remarks are needed here. The first

----

[1] Various conceptions of the researched object were considered by Roman Ingarden in the article "Betrachtungen
zum Problem der Obiektivität, in: "Zeitschrift für phi-
losophische Forschung" Bd. 2, 21, 1967.

concerns the term "subjective element", which is often
variously and erroneously interpreted. The objection is
sometimes raised in discussions that language is something objective. Of course, not only language, but also
e.g. measurement, measuring instrument, and finally the
whole "rest of the world" in relation to the cognized
subject is something objective. It is only that in a
definite cognitive process these elements can occur in
such a subjective epistemological characterization as was
presented below. The other remark concerns the fact that
an epistemological insight into the cognitive situation
of the object does not resolve (and no such attempts are
being made) the question of the ontic status of acquired
properties in the sense mentioned above. This means that
ontological problems connected with the subjective element
are without the scope of such considerations. For example,
the question is not resolved here of the ontological
status of the measuring instruments, of the "rest of the
world", or of language as it is in itself. Nor is the
question decided of the ontic character of the principle
of indeterminacy. What is important and interesting is
the role and epistemological characteristics of the principle of indeterminacy, and not the ontological status of
indeterminacy as such. Therefore, excluded from these
considerations are solutions in favour of or against e.g.
the Copenhagen interpretation of quantum physics, or any
other interpretation regarding the question of the ontological status.

After these introductory remarks, it is time to proceed to the above-mentioned more specific questions and to
the answers that can be given to them on the basis of the
source material mentioned.

I

Why have the character, role, and the very concept of experiment changed in 20th-century science?

In connection with this question, it is worth recalling that in the traditional philosophy of science and natural science experience meant something accessible to the senses, something "visible", something given to the senses or their technical extensions. It was believed that it was possible to derive from experience by logical means the fundamental concepts and laws of science, that scientific procedure consisted in collecting data and systematica-ly ordering them, that theory was built of data (of that which was given in experience) with the aid of the laws of logic. Connected with such a concept of experience was the commonsensical conception of the scientific fact.

The fathers of new physics share the view that given such a conception of experience actual scientific procedure becomes incomprehensible, since the multifarious aspects of the role and concept of experience in science were incorrectly presented. In new science, a different interpretation of experience proved necessary, since its character is more complicated. It has been expanded, and so it was conceived of too narrowly. It is also being asserted that some qualitative change has occurred in experience. Examined below are three selected aspects of this change.

(1) Of inestimable epistemological value is Heisenberg's laboratory observation that in every experiential situation in quantum physics a division into "the object" and "the rest of the world" constitutes the starting point. This state of affairs has vast consequences, inter

alia: (a) the very procedure of measurement is involved
in a complicated tangle of theoretical questions;
(b) that which is given in the experience is accessible
through the measuring instrument (Eddington says of this
situation that that which is accessible to the researcher
in the laboratory cannot be freed from reading the positions of the pointers; (c) the measuring instrument
directly interferes in the experiental situation. Therefore, Heisenberg so strongly emphasized the fact that
only such an instrument is a measuring instrument which
is in direct contact with the "rest of the world" and
when this instrument and the observer interact. Let us
add that the measuring instrument is constructed by the
observer on the basis of the existing knowledge. And if
so, the experience reveals that which the observer wants
to know (inter alia, the observer asks questions through
the measuring instrument in a definite manner). This
means that experimental apparatus (and through it, the
"rest of the world") acts upon the object researched.
Consequently, it means that the existing knowledge which
the observer has about the "rest of the worldj acts upon
that which is revealed in the experience. For in the
experiental situation in quantum physics, that which
occurs during the act of observation means actualization
of one of the possible events that the probability function
describes as an experimental situation at the moment of
measurement, and thus in an experiment carried out by
definite means.

(2) The second aspect of the complicated character of
the experience in contemporary physics is connected with
the fact that a special role in this physics is played by
some classical notions. The point here is not only that
they serve to express the experience but also that they

constitute a condition of the observation of atomic events. Such notions as time, space, or causality, shaped in the course of evolution of human cognition, are indirectly linked with the experience and have been inherited by contemporary physics (in this sense, they are frequently regarded as being of an aprioristic character). They constitute a part of the system of physics and serve to define the remaining part of the universe in describing the researched object. So classical notions, as an inalienable aprioristic element, condition the observation of events in the atomic world.

(3) The third aspect of the complexity of construction of the experience in contemporary physical sciences is in some way linked with the previous one. What is meant here is the theoretical element of the experience, the element that intervenes in every attempt at understanding the world as an inalienable integral component. The presence of the theoretical element in the experience is stressed, amongst others, by Heisenberg in his conception of so-called closed theories. Especially interesting are some of the characteristics of this theory, namely: (a) it provides a description of nature with limited validity; (b) the notions employed by this theory have some meaning in the world of phenomena (as Heisenberg says, they must be enrooted in the experience); (c) this theory is a part of the language of natural science.

Altogether, such would be the answer to the questions about a change of the character, role, and the very concept of experience in contemporary science as compared with the existing knowledge, that is classical physics.

II

Theses about the visual and non-visual nature of science are generally known in the philosophy of science. The former, linked with classical physics and the mechanistic outlook on the world, implied the methodological postulate of the visual character of science. Visualness was considered to be an indispensable feature of scientific cognition, and physics was regarded as a description of phenomena accessible to the senses, based on experience and mathematics.

After the breakthrough caused by the relativity theory and following the quantum revolution, the question of visualness of science reemerged in discussions amongst scholars. The very postulate of visualness was judged critically. It was observed that science offers notions which can neither be represented in pictures, nor imagined, nor described. The researchers themselves noticed that the value of their thought does not depend on its visual character: physics had entered areas inacessible to the senses; the abstractness of science had increased.

Without denying the above theses in view of the problem under consideration here, one should emphasize another thesis which exposes a characteristic of postrevolutionary science insufficiently underlined to date. For contemporary science cannot wholly reject visualness, and lack of visualness, even though it does not lower the value of theory, poses a serious obstacle: it forces scholars to make an additional effort which is an authentic, creative cognitive effort.

In connection with the above, an answer to the following specific question will be presented below: why does such a paradoxical situation exist in science?

Although science has gone beyond the framework of sense experience, maintaining a certain tie with it is necessary for epistemological reasons that express themselves in methodological rigours. Especially significant from this point of view is the requirement of experimental verification of propositions, a requirement which, if not met, may lead to the accusation of cognitive worthlessness. From a philosophical point of view, it is especially important that by the creative effort of scholars and at a certain cognitive price, the aforementioned paradoxical situation in science is being eliminated: it is an effort aimed at visualizing contemporary science.

This compromise between the dissimilarity of phenomena of the atomic world and the necessity to express them in the language of classical physics is well illustrated by Heisenberg's indeterminacy principle, for it reveals (from this point of view) cognitive losses resulting from such a compromise. It demonstrates that exact measurement of the position in the world of atomic phenomena is possible if we simultaneously relinquish equally exact measurement of the velocity of the atom particle. Thus the exact knowledge of one of the parameters, the so-called coupled quantities, is entangled in the consequences of relations of indeterminacy. At this price, however, a certain degree of visualness of contemporary science is possible: it is possible to the extent to which measurement is possible.

Similarly, this compromise has found expression in Bohr's complementarity principle, which makes it possible to recognize as correct various visual pictures of the atomic system, even if they are mutually exclusive. Although the mathematical schema of quantum theory conceives in a non-contradictory and comprehensive manner experimental data supplied by these mutually exclusive descriptions, nevertheless science is forced to reach for visual pictures, even though it does this at a certain price, namely, visual descriptions of atomic processes have the value only of incomplete analogies.

Concluding, it should be said that science, doomed to visualness as it were, can achieve it in its development only by incorporating the existing theoretical knowledge which governs experience.

From the point of view of the problem under consideration, a successive specific question can be posed.

III

Why is the description of the state of the system in contemporary physics (e.g. as regards microphysical phenomena) characterized by certain peculiarities?

One can speak of peculiarities only comparatively, in this case - in comparison with such a description which did not reveal these peculiarities. Therefore, it seems appropriate to recall here in brief the characteristics of the classical (mechanistic) description of the state of the system.

In classical physics, description means determining the state of the material system at any time on the basis of the laws of physics and the initial state of this system. Knowledge of the initial state of the system is provided by the description of coordinates (e.g. position and momentum established by way of measurement).

From a methodological point of view, the same rules of procedure are in force in contemporary physics. However, the initial state of the system is not determinable in the sense mentioned above. In effect, the final description

of the system as understood by contemporary physical
sciences violates the canons of description adopted in
classical physics. It is a denial of the classical description in science: it strives for something (and achie-
ves it) that in principle was eliminated by the classical
description. One can even say that it is the reverse procedure in relation to the classical description, for in
the latter, during the description of the state of the
material system, this description is concretized by means
of definite parameters, whereas here to describe the state
of the material system means to reveal its parametric indeterminacy. To the question of why it is so, one could
briefly answer: because what constitutes the obstacle is
a set of factors whose influence upon the course of phenomena in the macroworld could be disregarded by physics.
Such an answer, however, tells us little about the problems under consideration, and a more detailed analysis of
this shortcoming would be needed. Examined below are only
its selected moments, of special interest to us here.

Special difficulties are posed by the language of
description. Not only the measuring instrument is subject
to classical description. The description of phenomena,
experiments, and results is expressed in the language of
classical physics. The problematicity and indefiniteness
of the notions of this language with respect to the description of microparticles causes ambiguities in the description (they can be eliminated only by speaking about
these objects in the language of mathematics).

Let us note that the specific paradox of the situation
of description in microphysics consists in the necessity
of extrapolation of notions of classical physics (that is,
notions serving to describe objects behaving classically)
to the field of objects, of which one cannot say that they

behave classically. Thus, the necessity of this extrapolation does not result from the classical behaviour of microobjects, but is linked with the mode and method of conduc-
ting research, is epistemologically enforced. Thus the impossibility of freeing oneself from the existing knowledge
(the old theory) is an epistemological necessity.

Indeterminacy occurring in the description of micro-
phenomena is caused by such a cognitive situation in which
the researcher, in describing an experiment, must use notions of the language of classical physics, that is to say,
not fully adequate notions. However, the division into
"objects" and "the rest of the world", necessary in the
experimental situation in quantum physics, is conditioned
by the state of the observer's knowledge (that is, the
existing knowledge).

Indeterminacy as a certain kind of knowledge does not
concern the microobject itself, but the knowledge of it.
Since indeterminacy is connected with the necessity to
use, in describing microobjects, notions developed for the
description of macroobjects, it can be concluded that we
do not possess such additional knowledge that would tell
us whether the kind of quantities (such as momentum,
velocity, position, etc.), worked out in the course of
cognition of macroobjects, is appropriate for microobjects. In other words, we do not know whether such
extrapolation is valid for reasons other than epistemological necessity resulting from the fact that without
using visual notions of the existing theory one cannot
describe the experience.

From such a point of view, indeterminacy is the epistemological price that has to be paid for acquiring new
knowledge, considering that one already previously posessed

some knowledge, gained in one way or another. The researcher cannot free himself from this manner of gaining knowledge, he can at most gain knowledge about this fact. The inevitability of including the old (existing) knowledge in the acquisition of new knowledge is illustrated by what Heisenberg called a division into "objects" and "the rest of the world". This "rest of the world" influences (through the measuring instrument) the object researched, that is to say, the knowledge of the rest of the world is present in the knowledge of the object.

Summing up, one should stress that the experimenter's possibility to choose an aspect of nature at the expense of resigning from another aspect - a possibility revealed by both the indeterminacy principle and seeming contradictions in visual pictures of undulatory-corpuscular dualism (it is seeming because it occurs only on the level of these visual pictures) - not only demonstrates the inadequacy of the notions used in the description, their inapplicability in describing the phenomena of the macroworld, but above all reveals the necessity of employing notions of classical physics in describing results of observations. The situation of description of macrophysical phenomena distinctly shows the relation between the new knowledge and the existing one.

IV

After the 20th-century breakthroughs in physical sciences, the traditional ideal of the objective description of the world has lost any significance. The cognitive situation of contemporary physics does not make possible the actualization of this ideal, according to which the role of the so-called observer boiled down to passive observation of nature, while the perceived nature appeared to be something entirely independent from, external to the observer. Also, the Cartesian division into res extensa and res cogitans was still valid.

An in-depth analysis of the post-revolutionary cognitive situation in which physical knowledge is achieved leads to the thesis about an increase of the participation of the subjective element and, accordingly, a growth of difficulties and of the cognitive effort in overcoming this subjectivity. In connection with this thesis and the basic theme of this paper, the next question is the following: why is the participation of the subjective element in scientific research increasing?

As stated above, the complexity of the description of the atomic system consists, inter alia, in that this description not only concerns the object, but of necessity the description of the rest of the world participates in this description (in the sense explained above). The distinguishing of the object alone is an act of selection. Selection assumes something being selected from something. Selection depends on being equipped with means of acquiring knowledge and in this sense is subjective. Scientific description concerns so-called good observation, that is, selective observation. Therefore, a selective test is added to physical knowledge. The very division into "object" and "the rest of the world" as a consequence of scientific method is in a way arbitrary and depends on the existing knowledge.

Still valid in contemporary science is the postulate of describing the experience (including the measuring instrument) in terms of classical physics. The instrument is constructed by the observer on the basis of the existing

knowledge. What constitutes the subjective element in describing microevents is language as a means which the researcher possesses, in which he formulates questions in a
definite manner and, in this sense, he views nature in the
"given" manner.

As stated earlier, measurement interferes in the experimentally intended course of microprocesses. Although
measurement in the field of application of the quantum
theory is also something objective, its intervention in
the course of a phenomenon consists in cocreating physical
properties of the atomic system, and this participation
cannot be isolated in description. Thus, knowledge about
the working of instruments, expressed in classical terms,
cannot be separated from description.

Such description no longer meets the traditional,
classical requirements of objectivity in the sense of passive registration of what is taking place, for precisely
the act of registration influences the knowledge contained
in description. The probability function comprises contents
independent from the observer as well as subjective
elements (e.g. concerning the knowledge of the system,
and this knowledge can vary from one observer to another).
The subjectiveness of this knowledge stems also from its
limitedness (the observer's knowledge about the "rest of
the world" is incomplete). Introducing the observer, i.e.
the registrant (a man or apparatus that registers) is
necessary because it is necessary to register a material-
izing possibility in the course of measurement. Partici-
pating in this act is also the knowledge contributed by
the observer.

Summing up, with the question posed in this section in
view, it should be said that analysis of such a research
peculiarity as e.g. describing the state of the system of

the microparticle, conducted from the viewpoint of the
subjective element that is present in the knowledge being
acquired, reveals, generally speaking, the irremovable
necessity of the presence of the existing knowledge in
achieving new knowledge, or participation of the existing
theory in research conducted within the framework of new
theory.

V

Below the title question will be examined in the light
of indeterminism and causality. As known, quantum physics,
by denying the unequivocal character and continuity of
physical processes, undermines the principle of causality
as an absolutely binding principle in science and indicates the acasual nature of phenomena of the atomic
world.

In classical physics, causality is linked with the
possibility of unequivocal prediction. On this deterministic causality, the general postulate of determinism is
founded, according to which the absolute necessity of the
run of phenomena prevails in the world.

Quantum mechanics provides grounds for the conclusion
that in the atomic world there do not exist close causal
ties as understood by classical physics. In the new
physics, connected with the specificity of contents is
the incompletely defined knowledge of the atomic system:
descriptive quantities, e.g. electron, can be known only
to a limited extent, which appropriately restricts the
possibility of prediction. Generally speaking, even though
the future is not fully independent from the past, it
nevertheless is not completely determined by the past.
Quantum mechanics is unable to establish what will happen

with the atom at the next moment, for it can change its
state in various ways, and therefore its future is not
unequivocally determined. In microscopic processes, time
sequence raises doubts, for in the quantum theory the time-
space structure ceases to perform its fundamental function,
and it becomes difficult to reconcile the time-space description of the object with causality. Under the same conditions, various possibilities can materialize. Thus, the
principle of causality does not apply to phenomena in the
atomic world, and the postulate of determinism fails.

Let us note, however, that the epistemological characteristics of indeterminism in present-day physical sciences
warrant perceiving the source of the departure from the
deterministic interpretation of phenomena in the aforementioned limitations imposed upon scientific description.
For example, (a) the equivocal definition of transformations of the atom in time during the process of disintegra-
tion of radioactive substance indicates the problematic
character of time sequence and consequently indicates time
indeterminism; (b) the limited precision of simultaneous
measurement of coupled quantities describing electron, and
thus the limited possibility of prediction, indicates indeterminism connected with Heisenberg's principle of indeterminacy; (c) the fact that the wave equation does not
permit to predict the change of the measured system occuring during the measurement processes, nor does it permit
to define the state of the measured system after the
measurement, indicates measurement indeterminism.

Limitations imposed upon description, which are the
cause of the incomplete knowledge of atomic systems, force
us to abandon the classically deterministic interpretation
of physical phenomena, which limits causal description in
atomic physics. Analyses of the peculiarities of scientific

description in contemporary physical sciences (presented here earlier) each time showed that we are entirely doomed to the existing knowledge, to the participation of classical concepts in the new knowledge.

In this sense, the assertion seems justified that contemporary physics has not freed itself from the category of causality (nor has it freed itself from certain classical concepts). Therefore, being aware of this fact, some physicists, in their reflection on science, prefer to speak of a limitation or the necessity of modification of the category of causality (e.g. Cz. Białobrzeski speaks of multivocal causality, i.e. deterministic and indeterministic).

It appears, however, that not only in this sense we cannot wholly reject the category of causality. For if the specificity of description of the state of the microobject forces us to modify this category, then having in mind the conclusions and characteristics of scientific description, we can arrive at a seemingly paradoxical thesis that it was not so much indeterminism that removed determinism from science, but rather determinism led to conceiving of phenomena in an indeterministic spirit. One should presume that such an idea must have guided Eddinton, when he wrote that the rejection of determinism in quantum physics is not abdication of scientific method, but rather a fruit of scientific method.

Summing up, let us recall the conclusions drawn here earlier. They boil down to the thesis about the inevitable presence of the existing knowledge in gaining new knowledge; its active interference in the cognitive process (e.g. in measurement); its presence through the inclusion of the knowledge of measuring instruments expressed in classical terms; the existing knowledge conditions the

distinguishing of so-called good observation; and any
sudden change of the function of the state reflects the
incompleteness of the observer's knowledge about the
system. These epistemological considerations, this impossibility to free oneself from the existing knowledge
in achieving new knowledge, are responsible for the limitations imposed upon description in atomic physics. This
description necessitates rejecting deterministic interpretation of natural phenomena and makes necessary reinterpretation of the law of causality in new physics.
In the final conclusion, these considerations again point
to the relation between old and new knowledge.

VI

As a consequence of interrelations examined here, a
number of other questions in the field of the philosophy
of science must undergo an appropriate reinterpretation.

It should be remarked, for example, that rejection of
the classically deterministic interpretation of natural
phenomena influences the relation between facts and
theory in such a manner that the dividing line between
theory and sentences describing facts is blurred. What
further influences this fact is the conceiving of the
experience as a complex construction and the presence of
the theoretical element as well as interference of the
concepts of the existing knowledge (as stated earlier).
In effect, it is impossible to separate description from
the whole theoretical equipment, that is to say, the so-
called pure description is impossible.

VII

Another example is offered by a difference in conceiving of the relation obtaining between such notions as part and whole. The indeterminism of contemporary physics and the necessity of modification of the concept of causality require a manner of explanation of natural phenomena different from that applied in analyses founded on a mechanistic vision of nature. This concerns, above all, a different interpretation of the relation between such notions as part and whole. According to present-day physics, the whole is no longer a set of parts, machinery. The whole is something more than the total of parts, it defines their meaning. The laws governing it cannot be derived from the knowledge of parts, and it is to such laws that the parts of the whole are subject. In quantum mechanics, they take the form of mathematical formulations, and it is possible to derive from them the characteristics of atoms - wholes, which they govern. The different conception of the relation: part - whole, leads to replacing the classical procedures with analysis of the procedure of structuring, or such mental operations which always make part dependent on the full set of parts as a structured whole. Therefore, the importance of parts is dependent on the whole system. Such a structuring procedure can also be called structural analysis. The knowledge provided by physical sciences is a structural knowledge. For reasons of basic importance for post-revolutionary science, the very notion of structure is linked with mathematical formulae, properties of the mathematical symbol, mathematical description of the group, in other words, with the mathematical conception of structure.

VIII

The authors of quantum theory were keenly aware of
the uselessness of classical logic for this theory and
considered the possibility of applying another logic
(notably Heisenberg and Weizsäcker).

Classical Aristotelian logic, which is applied both in
everyday life and in classical physics (both colloquial
language and the language of traditional physical theories
are subject to it) becomes useless in quantum theory, if
the language of this theory is to be unequivocal, and
this happens when it takes the form of an artificial
mathematical language. In a language adapted to the mathematical formalism of quantum theory, classical Aristotelian logic would have to be replaced by quantum logic,
from which would be absent the basic axiom of the
Aristotelian formal system: the principle of excluded
middle.

Quantum logic would permit intermediate situations -
such situations in which it has not been decided whether
a sentence is true or false. And a sentence reflecting an
intermediate value cannot be expressed in colloquial
language (according to Weizsäcker, it is a sentence comple-
mentary with respect to simple alternative sentences).
Heisenberg illustrates this dissimilarity of the logic
of thinking by the example of a box divided into two
halves by a partition with an opening in it: according to
quantum theory, it is logically wrong to believe that the
atom is either in the left or the right half of the box;
there are other possibilities.

From the point of view of our enquiries, especially
interesting are the considerations of the fathers of

quantum theory on the possibility of applying such quantum
logic. For the language of the mathematical formulae of
quantum theory requires a logic different from the classical one, but at the same time there are difficulties
with applying this new logic (if it is possible at all).
This paradoxical situation (the need to apply a logic
other than Aristotelian and the impossibility to apply it
because it is different from Aristotelian) brings to mind
the earlier considered dilemma of new physics (the need
to depart from the notions of the classical language and
at the same time the necessity to employ these notions).
But this is not the only similarity of situations which
can be described as paradoxical: for the problem is
always the same, it is only considered from different
angles. It is because quantum physics cannot free itself
from the classical language in the description of the
state of the system that it cannot distance itself from
classical logic.

For the philosopher of science (and the more so for
the epistemologist), greatly instructive are the manners
of interpretation supplied by those philosophizing
scholars and concerning the relation of quantum logic to
classical logic. For this interpretation is marked by the
same spirit in which they interpret the relation of quantum physics to classical physics. For example, quantum
logic is an "extended" logic, the border case of which is
classical ("ordinary") logic. What is more, in the language of the latter, a description of quantum logic and
its structure is possible.

So once again - this time on the plane of logical
problems - it is necessary to repeat that contemporary
science itself makes it necessary to take into account

and examine the situation of diachronic change connected
with the scientific breakthrough, when new theory replaces
the old one.

IX

As stated in the introduction, the last word will also
concern the question of diachrony - but this time critically. We will again begin with a question. Why are those
conceptions of the development of science which refer to
the correspondence principle so helpless in the face of
criticism? This criticism, though extensive, in fact
boils down to the charge of emptiness behind the formula
of this principle. A short answer to this question, in the
light of the enquiries presented here, is the following.
Because the formal record of this principle does not
reflect (if it does not lose) the whole richness of the
epistemological imperative of attachment to the existing
knowledge (i.e. classical Newtonian physics) that can be
revealed in the anatomy of the cognitive situation of new
science (non-classical physics). The advocates of the correspondence principle in the domain of the philosophy of
science have never taken the trouble to look at the
situation in such a manner as to see that the very formula
of the correspondence principle is something secondary with
respect to this situation, is a certain response to this
situation (most frequently, its formal expression). The
supporters of the correspondence principle, in defending
it (though it is only alleged defence), limit themselves
to stating that this principle is functioning in science,
in its history. Such a defence is very weak and therefore
helpless in the face of accusations.

True, it is a critical remark,but luckily the last one.

THE ANTINEWTONIAN CONCEPT OF THE OBSERVER

M. ZABIEROWSKI

Technical University of Wrocław, Wrocław
Wybrzeże Wyspiańskiego 27, 50-370 Poland

ABSTRACT

The antinewtonian global category of perceiving
Man, proposed by Wheeler, is rediscussed in the
new way. This sensualistic construction leads
to further consequences with respect to the anthropic modification of the objective physical
cognition.

1. INTRODUCTION

Depersonalized physics was to be a pattern for the
objective and precise cognition: the laws of physics and
fundamental constants would be separated from the elements
of consciousness and the existence of the physical world
would not be connected with the existence of a cognizing
subject /Man/; physical laws invariable in relation to
the transformation of time and place would be obligatory
/valid/ irrespective of the fact whether Man exists or
not.

Deviations from the Newtonian vision of the objective
science are treated as a curiosity. For the first time
these deviations appeared in quantum physics but only in
the attempts to interpret its formalism. In the presence
of the whole range of possible interpretations these de-

viations evoked neither controversions nor great hopes
that they would be quickly confirmed or strengthened.
The need to find non-psychologistic interpretations of
quantum physics is to be seen in the following thesis:
"In the whole Copenhagean interpretation /as well as in
the others/ the meaning of a measured object, a device
and an observer is not clear. There should exist such
a conception of measurement structure in which mutual
relationship between these elements would be defined.
Only in such an ill-ordered conception of measurement
which emerges from the Copenhagean interpretation the mis-
conception of the Schroedinger´s cat could arise. In the
authors´s opinion the inclusion of the psychological
elements to the determination of the state of the tested
object, i.e. in the case of the statement about cat´s
death or its good health reminds the inclusion of the
consciousness elements into the drawing of the round in
the sport contests"[1].

In this paper we want to present an example of another
"odd" approach in cosmology. The interference of a cognizing subject with the structure of the physical Universe
constitutes a significant novelty in modern cosmological
considerations. The formulation of so-called anthropic
principle /AP/ in cosmology [2], which is in a few versions,
is a something of a turning point in the very cosmology
and philosophy.

The disputes over the physical and chemical interpretation of life and consciousness are well known. The
problem of the interpretation of life processes only on
the basis of the unquestionable achievements of science
has not lived to see realization. The weak results of
this reductionistic approach to the life are often
accounted for the fact that biologists encounter a high
degree of complications. Moreover, the enthusiasts of this

point of view connect scientific method with physics or
chemistry fighting with every possible manifestation of
anthropocentrism. The conceptions that physical sciences
cannot do without non-physical treatments would be an un-
pleasant defeat for these optimists. While the philosophical systems have acknowledged the irreducibility of
consciousness to the physical-chemical world, the
strength of these reductionists derives from the rejection of "unreliable" philosophy for the benefit of the
only right science model, which physics would be the best
example of.

Although the whole effort of these biologists, who
wish to demonstrate unsuitability of the reductionistic
vision, leads to clearly declarative theses about the
existence of processes specific only for life, it is
worth to notice that the "model" science about the world,
i.e. physics, is shaken not because of the non-adjustment
of some significant physical parameters but at the very
methodological assumption that physics must be depersonalized science, devoided of any psychologism. Psychologism in physics sounds like an unfortunate attempt of
doctrinal disposal of the "Cartesian method".

Usually the conceptions have been adopted that
a scientist is not able to evaluate the rightness of the
thesis about the possibility to bring /reduce/ life to
physics. Meanwhile it seems that modern cosmology suggests a special link between the physical processes and
the existence of Man. Physics has never been demanded
to surpass the material reality: the acknowledgement of
this possibility would be the expression of metaphysics
but not of the scientific standpoint. Cosmologists reach
constantly beyond the territory of intellectual exploration marked for them in terms of Newtonian and Cartesian
methodological convention. Namely, in the present paper

we shall deal with these visions which have recently
appeared in relativistic cosmology and which treat the
Universe evolution as the realization of some kind of
mission, obligation to life. Speaking about the mission
the problem of a messenger and a dispatcher appears.
In this article, we shall not give an exhaustive answer
to the question formulated in this way, though this
problem marks the perspectives of our considerations.

2. THE WEAK FORMULATION OF THE ANTHROPIC PRINCIPLE

Whitrow [3] shows that Einstein's equations seem to
forget about the fundamental truth that "great" Universe
is a house /a dwelling-place/ of all living organisms.
That is the structure and the evolution of the Universe
can only be fully understood when besides the very
Einstein's equations the theses of biological character
are taken into account. Whitrow [4] believes that not only
the size of the Universe is a free parameter, but also
the three-dimensionality of space is connected with the
fact that space parameters are neither unrestricted nor
"external" in relation to human existence. Independently
of Whitrow this problem has been undertaken in cosmology
and its essence is expressed in the anthropic principle
/AP/.

The weak AP derives from Dicke's work [5], in which he
shows that in many respects the human situation in the
Universe is by all accounts favoured, that is Man cannot
consider the Universe as an accidental configuration of
matter, fundamental constants and physical parameters.
The standpoint which makes the Universe parameters
dependent upon the subject is inadmissible from the point
of view of the puristic principle which separates the
world of the cognizing subject from the physical Universe.

Making, by Dicke, the cosmology subjective was a consequence of the following deductive chain: life comes into being in galaxies, thus there must be sufficiently many stars supplying heat; this means that the number of years which have passed since the primeval big bang cannot be very great as the stars would start to wane; consequently, the time scale established by Hubble's constant /the Universe expansion rate/ cannot considerably exceed the average time necessary to synthesize heavier elements which are indispensable to the origination of life; however this period of time cannot be much smaller than 10^{10} years /estimated Universe age/since then the theory of life appearance as a result of the origination first of heavy elements and then chemical compounds would not be valid. Therefore in this sequence of reasoning there appears clearly an assumption that heavy elements came into being as the consequence of hydrogen "combustion" /nucleosynthesis/ and then they became a decisive factors in the process of the life origination. Even the simple considerations referring to electromagnetic radiation scattering and to hydrogen combustion in stars allow to establish the number of years which have passed since big bang, to approximately 10^{10} years. Therefore since a human being is already in the Universe, its age and numerical values of the other astrophysical and cosmological parameters are not accidental but they are defined just by the fact of Man's existence.

In the presented approach to the Universe evolution , the unshakable thesis of the life appearance is the starting point of every reasoning. The evolutional origin of life is taken for granted. The aim of such reasoning would be to establish the parameters which do not result directly from physical theories, e.g. the age of the Universe counted from the big bang.

The reciprocal of Hubble's constant [6] 1/H established
on the basis of the fringe shifts in galaxy spectra is in
agreement with the data from heliophysics and astrophysics as far as the order of value is concerned.

Heuristics of the programme formulated in this way
can be expressed more or less precisely in the following
way: when establishing numerical values of the cosmological parameters adopt the rightness of laws known from
physics, do not change these laws /do not follow e.g.
Jordan, Dirac, Tifft, Arp, Hoyle, Bondi, Narlikar [7]/,
but derive values of the parameters of Universe from the
physical laws and the theory of the evolutional origin of
life, apply the method of co-ordination between physics
and biological thesis about the life origination.

The weak AP helps to establish the Universe parameters
such as: density, age, size, degree of isotropy and
homogeneity of the substratum, curvature, other parameters
connected with grouping of matter and the derivatives of
these quantities. It says about the only Universe inhabit-
ed by a human being /it does not introduce the notion of
the Universe "ensemble"/. The weak AP affirms the evolutional model of the Universe but it does not support
the theory of Hoyle-Bondi-Gold-Narlikar, according to
which the Universe would not have the time beginning.

3. THE STRONG ANTHROPIC PRINCIPLE

The necessity of the subject appearance is not the
essence of AP. The thesis that the cognizing subject must
come into being has a different meaning than the considerations starting with the words: "Since life is a fact
then ..." etc. Each reasoning chain beginning with the
assumption of the observer's presence in the Universe
/as in the weak AP/ and leading to the establishing

certain properties of the Universe is to some extent a
post factum explanation.

The strong version of AP says that the Universe
parameters should just be such that life has to come into
existence. The human consciousness can be then regarded
as a self-consciousness of the Universe, i.e. as the
ability to present the knowledge about itself. According
to the weak AP, Man can conclude about the Universe
making use of the fact of its existence but he cannot
present himself as the aim to which the Universe endeavours, as a being that must come into being some time at
the Universe will, as it is required by the strong AP.
According to the latters the thinking is the act of the
Universe manifestation. The weak AP entitles to estimate
the cosmological parameters by the terms of the subject
existence. In the strong AP the cognizing subject is an
inseparable Universe attribute: the Universe and thinking
cannot be separated each other. If the Universe could
speak it would repeat after Descartes: "cogito ergo
sum", whereas Carter suggests a paraphrase: "Cogito ergo
mundus talis est" [8].

The strong AP attracted the attention of physicists
as its adoption was tantamount with the acknowledgement
that the existence of the worlds, which were different
from ours, belonged to its prediction. The existence of
the other worlds, suggested earlier by Everett [9] as an
interpretation of quantum mechanics could not defend
itself from the argument that this existence was antici-
pated only by his unique approach. The comments on Everett
interpretation recall the famous remark about a research
paper once made by Wolfgang Pauli:"It is not right. It is
not even wrong" or the Richard Feynman remark: "When
playing Russian roulette the fact that the first shot got
off safely is little comfort for the next. There is

nothing much so wrong with this as believing the answer!".

The interpretation of the state vector in Hilbert's space as a superposition of the states representing mutually isolated worlds found an ally just in the strong formulation of AP. The paradox of the Schroedinger's "half-dead" cat was solved in this interpretation by the admittance of two cats: a dead one and an alive one. Any of these cats was never observed by the other one because the monadologic construction excluded any "windows" in these two worlds /the world in which the cat lived and the world in which the cat was dead/. The lack of these "windows" implied the possibility of paradoxical acknowledgement of wave function which described the half-dead cat.

The strong version of AP indicates the possibility to introduce to our knowledge many separate worlds from which only one can be observed by us. It refers directly to the dendrological interpretation of the state vector of the the universe [10) /the universe consists of the our Universe among others/. All the dendrite branches are equally actual from the dendrite's point of view /super-observer/ and in this sense the branch of our Universe is not favoured though Man cannot take part in the evolution taking place in the remaining branches. Each branch represents the different world of physical constants and boundary conditions. Both the parameters of big bang and the values of physical constants are defined ab initio /the values of these parameters and constants are not derived from the fundamental theory/. We have no empirical knowledge about these physical quantities in the other branches but it can be shown why their values in our world are as they are.

The strong favours the position of Man and in this way it is,to some extent, contrary to the Copernican

principle. The idea of many worlds is an attempt of
reconcilation between these two principles: in the face
of the possibility of life origination in very similar
conditions to these of our worlds not Man is favoured but
rather the whole class of subjects that can be "cultivat-
ed" in the other - unobservable by us - universes.

The set of the possible universes comprises also such
a branch in which the subject formulates the answer to
the question: what the Universe is? It means that if it
is possible to ask about the role of the subject in our
Universe it is also possible to transfer this question
to the level of the universe ensemble. The negation of
the orthodox Copernican principle in our Universe is not
tantamount with its rejection on the dendrite level: our
Universe is not the only one.

4. THE ANTHROPIC PRINCIPLE AS THE SUPPLEMENT OF THE PRINCIPLE OF THE SPONTANUOUS UNIVERSE ORIGIN

Zeldovitch [11] trying to explain the origin of our
Universe arrives at the conclusion that AP helps to find
the explanation to this problem [12]. According to Zeldovitch the world came into being in singularity - it did
not exist before. Zeldovich does not consider the question whether it is the matter of the creation from
nothing or of the topological separation of one world
from the other. The possibility of the existence of the
primeval empty Euclidian space, which all the possible
worlds would derive from, is permitted. This primeval
space is like an ether, the system existing before the
origination of our Universe. He makes use of the fact
that momentum energy tensor is defined in the case of De
Sitter's solution by the polarization of vacuum caused by
the curvature of the 4-dimensional Einsteinian space-time.

That is the reason why the question refers to the
"origination" of the universe with non-zero curvature
/we cannot treat the question of the creation of the
Euclidian world in the same manner/.

The primeval cosmological singularity is defined by
$a(0)=0$, where $a(0)$ is the scale factor $a(t)$ at the moment
$t=0$, i.e. at the moment of the creation of our Universe
which is closed and $(3+1)$-dimensional. The Universe expands from $a=0$ to $a_{min}=cH^{-1}$. For slightly bigger values
of the scale factor $a > a_{min}$ the evolution takes place
smoothly, according to the metric $ds^2=c^2dt^2-(H^{-1}$ ch $Ht)^2$
$[dr^2+\sin^2r\ (d\theta^2+\sin^2\theta d\varphi^2)]$, ch $Ht \sim \exp(Ht)$ for $t \gg H^{-1}$ and
the smooth connection takes place at the moment defined
by the Planck´s time 10^{-44}s. The quantum mechanical amplitude A of the spontaneous Universe origination as a result
of quantum fluctuation is defined by the expression $A =$
$= \int \exp\left(i \int_0^{t_{min}} LdVdt\right) d\omega$, where $\int_0^{t_{min}} LdVdt$ is the integral of the action L and the integration is expanded over
space-time with definite volume $dVdt$ while the integration with respect to ω passes through all scenarios of possible universes /each of these universes defines a dendrite branch/, $d\omega$ denotes an infinitely small difference
in the scenario of the created universe /so far it has
been made the use of the notion of the universe ensemble
and we shall come back to the question of the creation of
our Universe/. Only in the case of the closed universe
the amplitude A is non-zero. The closed universe is characterized by the zero value of total energy, momentum and
electrical charge thus there is not any breaking of the
conservation laws[13].

The expansion, established by De Sitter´s metric,
starts at the moment when the Universe reaches the size
given by the so-called Planck´s length. It ends rapidly
at the moment when the created plasma influences the phys-

ical enviroment so that the former state equation $p=-\varepsilon$ is
no longer valid /p is pressure and ε is density of energy/.
At that moment the state equation of the Universe is changed into the simple equation for the relativistic gas $p=$
$\varepsilon/3$,/the relativistic gas is characterized by temperature slightly lower than the so-called maximum temperature,
which was investigated by Sakharov and which equals approximately 10^{32}K/. In the model of the closed universe
the too short expansion phase of De Sitter´s type does
not lead to the significant cooling of gas , that is the
relict radiation has the temperature higher than 3K. In
this situation the long expansion phase of De Sitter´s
type taking place before the Friedmanian expansion occurs
the most desirable. These conclusions derive directly
from the possibility of easy scaling of physical quantities in the time function t, including temperature T in
the model of the closed Universe.

So far we have not pointed out the fundamental problem connected with the introduction of the explanation of
our Universe origin. The basic difficulty in these considerations is connected with the condition to ensure the
long expansion phase of De Sitter´s type, which excludes
big initial perturbations. However the principle of the
spontaneous Universe origin does not forejudge whether
the initial perturbations are big or small. For this reason Zeldovitch arrives at the conclusion that making use
of the anthropic principle is indispensable. It means
that the thesis about small initial perturbations belongs
to the AP predictions. Generally Zeldovitch believes in
gemnation of worlds,but the existence of our Universe is
guaranted by the relatively small perturbations of De
Sitter´s metric. He claims: since there exists Man in our
Universe then the perturbations of the assumed expansion
of the De Sitter´s type cannot be too big. If a universe

is devoided of a subject then the idea of the spontaneous
origination - permissible from the quantum theory point of
view - is sufficient without the need to refer to AP.

5. SENSUALISTIC ELEMENTS OF THE WHEELER´S INTERPRETATION OF THE ANTHROPIC PRINCIPLE

The cognitive values resulting from the quantum
mechanics especially from the Everett´s interpretation are
connected in a natural way with the possibility of the
"existence" of the ensemble of universes. In references
to the interpretation by von Neumann, who considered an
observer in quantum mechanics as the necessary condition
for the reality of a phenomenon, John Wheeler extrapolates
this thesis from the micro-world to cosmology. The Newtonian separate treating /isolating/ an object from an observer in epistemological sense is rejected in the
interpretation of quantum mechanics. The object is understood as a class of potentiality, and interference of the
observer is indispensable to define the real state of
object. Similarly, according to AP the existence of the
Universe cannot be separated from the cognizing subject.
Therefore it is not possible to isolate the physical object of an examination as independent from the perceiving
individual.

According to Wheeler we know easily what we experience but also there exists only that what we experience.
The Wheeler´s thesis belongs to the most extreme enunciations of sensualistic type that have ever appeared in
physics. The similar enunciations appeared in quantum
physics but have not upset the community of scientists
in the face of rivalry of different interpretations. In
cosmology the status of AP in Wheeler´s expression is exceptional. In some sense Wheeler is the opponent of using

exclusively mathematical formalisms in physics as in any
mathematical way it is not possible to express the thesis
that the subject is necessary for the Universe to come
into being. It does not longer resembles the Berkeley´s
judgement that colour exists only then when it is observed.
According to Wheeler the whole Universe exists under the
condition that the observer exists.

While the XVIIIth century sensualism identified the
existence in Nature with the perception of features of individual constituents the standpoint of Wheeler deals with
the relationship between the most globally understood universe and the cognizing subject. He does not stress the
significance of senses, thus he does not suggest any strat-
ification of the Universe to the same number of Universes
as Man has senses /for Berkeley the same table which is
touched and seen is already regarded as two tables/. Similarly, the repeated opening the eyes does not create the
world. Man is here a global category, therefore the construction of the world is not relativized to each perceiving subject and to his each state: the perception is generalized to the participation of the observer.

The Wheeler´s cosmological standpoint draws also
attention to the following Berkeley´s question: is the
Universe the perception only of people or also of God? If
in a way we can say about the Universe existence when
there was no Man, so who would then perceive the Universe?
The extraperceptive permanence of the world is meant here.
For Berkeley it is sufficient that in the interval between
subsequent human perceptions the things are "watched" by
God thus their existence is not a sequence of continuous
appearances and disappearances. Wheeler introduces the per-
ceptive being and the perceiving being: the Universe and
Man. But the limitation of the perceiving being only to
a man /or Man/ is insufficient to adopt the continuity of

the Universe, its constancy and its unity.

The Wheeler´s conception making the Universe subjective together with biological enunciation about the ephemerality of Man as a perceiving being seem to lead to further possible consequences of extranatural character. If all the dendrite branches of the ensemble of the universes are equally real, in whose mind does the idea of these universes exist? This time we touch the problem of the hierarchy of existence:it is possible to exist like our Universe does but also as the k-set of dendrite branches and as the ensemble of universes.

REFERENCES

1) Grabińska, Teresa et al.,"Heisenberg Relation of Uncertainty and Measurement Standards", Les Problèmes de Philosophie des Sciences Naturelles at de Philosophie de la Nature, 107, M. Lubański and Sz. Ślaga /eds./, Academy of Catholic Theology, Warsaw, 1984.

2) In the paper the author adopts the term "anthropic" after the English works in which the neologism "anthropic" is commonly and exclusively used to define the discussed principle. In some translations the term "anthropic" takes form of "anthropological" to express the transfer of human characteristics /e.g. preception, even will.../ to the "external" Nature.

3) Mascall, E., Christian Theology and Natural Science, Longmans, 43, London, 1956.

4) Whitrow, G., British Journal for the Philosophy of Science $\underline{6}$, 13 (1955).

5) Dicke, R. H., Nature /London/ $\underline{192}$,440 (1961).

6) The Hubble´s constant H establishes the rate of the Universe expansion. Its reciprocal 1/H is interpreted in the model of expanding Universe as the measure of time which has passed since the initial singularity

state. The moment of big bang is generally identified
with the time beginning of our Universe.

7) Eddington started a new trend called fundamentalism in
physics. According to his idea all the dimensionless
physical constants should be derived from cosmological
theory. Dirac and Jordan introduced the variability of
physical constants and in consequences different form
of the laws of conservation than it is generally adopted
/cf Lettere al Nuovo Cimento 26,349 (1979); Philosophical Studies-Polish Academy of Sci. (Studia Filozoficzne) 7,
105 (1980); Acta Physica Polonica B11, 471 (1980)/.

8) Carter, B., in: Confrontation of Cosmological Theories
with Observational Data, 294, M. S. Longair /ed./,
Copernicus Symposium, 10-12 September 1973, D. Reidel
Publ. Comp. , Dordrecht, 1974.

9) Everett, H., The Review of Modern Physics 29,454 (1957).

10) Wheeler, J. A., Rees, M. and Ruffini, Black Holes, Gra-
vitational Waves and Cosmology:An Introduction to Cur-
rent Research, Gordon and Breach Science Publishers,
New York-London-Paris, 1974.
Wheeler, J. A., Genesis and Observership, preprint -
Princeton, 1976.

11) Zeldovitch, J. B., Astrophysical Cosmology - Vatican
Study Week on Cosmology and Fundamental Physics,pp.575-
579, Pontificia Academia Scientiarum, H.Brueck,C.Coyne
and M. Longair /eds./, Specola Vaticane, 1982.

12) Worlds do not come into being with the same ease and the
probability depends on the adopted scenario.

13) The very low value of matter density adopted in the
seventies should not be regarded as reliable, cf   Zabierowski, M., Mem. Soc. Astr. Ital., 233-246 (1980);
Grabińska, T. and Zabierowski, M., Lett. Nuovo Cimento
28, 139 (1980), Nuovo Cimento 82B, 235 (1984), in: The
Cosmic Background Radiation and Fundamental Physics,89,
Società Italiana di Fisica, Roma - Bologna, 1985.

THE OPERATIONISM POSTULATE IN THE CLASSICAL KINEMATICS

TERESA GRABIŃSKA

Technical University of Wrocław,
Wybrzeże Stanisława Wyspiańskiego 27, POLAND

"Kto nas zakochał w swej sławy odbiciu...
Gdzie boskie, jeśli cesarskie w połowie?"
Z. BRONCEL, LITANIA DO NAJŚWIĘTSZEJ MARII

ABSTRACT

There are presented two attempts of operationism postulate realization in classical kinematics by Ives
and Kapuścik. In both approaches the non-observable
elements of theory turn out to be indispensable: the
absolute space-time /ether/ in the first case and the
fifth time-like dimension in the second one.

1. INTRODUCTION

The appearence of operationistic ideas in the methodology of empirical sciences was usually associated with
the famous work by Bridgman[1] which was regarded to be
the effect of the new physical theories which were new at
that time, i.e. special relativity /SR/ and quantum mechanics /QM/. In the first theory /SR/ the operational procedures of defining the kinematical quantities were to be
an integral part of it. In the second theory /QM/ the
measurement process influenced the description of physical
system state and all the quantum quantities, so-called
observables, were closely related to the measurement.
Hence all the physical concepts seemed to be determined

by means of empirical procedures. Operationism in physics
did not have only a descriptive character but primarily a
normative one.

The work by Bridgman opened also a new perspective in
the methodology of psychology, sociology, and other social
sciences[2]. In the times of neopositivism the operationism
found a stimulative climate but its very source should be
recognized earlier in the ninetienth century papers by
Mach and Poincaré, which were under the influence of positi
vistic conceptions. Here there appeares the question
whether the postulate of positivistic philosophy found
its full realization in the physical theories at the turn
of last century. The example of such a problem and its
solution was presented in the paper by Zahar[3]. He analysed
SR in respect to the consistency of its epistemological
assumptions with philosophical postulates of Mach who
undoubtly affected the outlooks of young Einstein. Zahar
set himself the task to investigate the question in a
maximum objective way apart from Einstein's own declara-
tions of his deep debt for Mach[4].

In many monographies[5] SR is considered as a pattern
of the operationistic physical theory. Zahar showed, however, that "Einstein's definition of time is compatible
with classical physics but also that all clasical theories
can be reformulated in terms of the time variable as
determined by Einstein's synchronisation convention"[6].
In this way Zahar strengthened the position of Mach[7] who
rejected SR as an example of the only apparent realization
of his own philosophy /Mach did not identify SR with his
philosophy/. Zahar distinguished explicitly between the
Machian positivistic philosophy and his intuitive methodology which essentially influenced the Einsteinian general
relativity. The first one expressed the operationistic
point of view on definitions of scientific concepts.

In the present paper we shall not repeat the Zahar's evidence that was sufficient to show the discrepancies of SR when one required the strict fulfilment of the operationism postulate. Similarly to Zahar we shall claim that only a weak form of this postulate was really assumed in SR. This form however did not allow to distinguish SR from the classical kinematics in the following sense: both in the cases of SR and classical kinematics the price in the form of absolute[8] elements of the theory was paid for the operationistic formulation of the theory. To show this we shall compare the Ives' kinematics[9] with SR and present an attempt to introduce the operationistic concepts of classical physics[10].

2. OPERATIONISM OF SR IN THE LIGHT OF THE IVES' KINEMATICS

In the lectures on SR Einstein devoted a lot of attention[11] to the measurement procedures of length and time. Although the basic invariant concept of SR was the 4-dimensional space-time interval the measurement methods which Einstein wrote about were deeply rooted in the measurement methods which were irrespective of time and space, similarly as in the Newtonian physics. At closer analysis it turned out that the procedure of distance measurement did not bring about any big practical difficulties if compared to the time interval determination. Two methods of clock synchronization were recommended in SR:

Procedure A

At the moment t_0 a light signal is sent in all directions from a chosen "geographical" network node /n=0/ in which a clock is set at $t_0=0$. At the moment t_n the signal reaches the node n of the network: the clock in this node should be set at $t_n=nl_0/c$, where

n=1,2,... . In this method the light postulate of SR
plays the essential role: it warrants the const-ancy
and universality of velocity c, that is, it assigns a
priori absolute velocity standard. The Procedure A
determining the time measurement by means of one
clock also is useful for the determination of light
velocity: a light signal is sent from a point O and
is reflected from the mirror placed at the distance d
from the point O; thus according to SR c=2d/t, where
t is the time interval read on the motionless clock
in the point O between the moment of signal emission
and the moment of return to the point O.

Procedure B

In the node n=0 two clocks are synchronized and then
one of them is shifted from this node to the subsequent nodes of spatial coordinate network according
to which each clock is set. In this case we deal with
a clock moving with a constant velocity w, the frequency of which will be slowed down as a result of
the movement in regard to the clock at rest. The
measurement procedure of this type leads to the clock
paradox known from textbooks of SR. However, if the
velocity of the clock motion is made infinitesimally
small then the deviation in synchronization using the
method B should be negligible if compared to the synchronization by the method A.

A peculiarity of the procedure A is strictly connected
with the light postulate: the impossibility of clock synchronization in general, when this postulate is not obligatory, is noticeable when one wants to make a synchronization by means of an arbitrary signal propagating in any
medium. Both of the methods A and B turn out to be utterly
inaffective since neither velocities of the synchronizing
signal nor the velocity of the synchronizing clock can be

established. The method B is not of any practical significance in the clock synchronization as the clock velocity w
cannot be determined if the space-time points have not
been earlier synchronized in a given inertial system. Thus
the method B is secondary to the synchronization procedure
A.

On one hand it seems that the light postulate was
necessary for the operationistic meaning of SR concepts as
it was understood by Einstein. On the other hand, however,
the same postulate was the main issue of SR criticism made
by Ives. He criticized it from the point of view of the
optician investigating the properties of light signal pro-
pagation and the state of empirical records which determi-
nated these properties, and from the position characteristic for a methodologist being of the opinion about conse-
quent operationism . As a methodologist Ives demanded
from physics /in this case - from kinematics/ that each
quantity occuring in a formula should have its operational
sense, i.e. apart from the theoretical postulates the
measurement procedures for appropriate physical quantities
should be expressed in a proper manner. According to Ives,
the Einstein's replacement of one absolute, i.e. the
 Newtonian ether by another one, i.e. the limited universal light velocity c made the operational understanding of
the occuring physical quantities worse.

For Ives the light postulate was not to be accepted
as there was no sufficient empirical record to acknowledge
that the light velocity was constant. In terrestrial
experiments determining the light velocity the procedure A
was applied while the procedure B with the moving clock
was preferred in SR. According to Ives these two methods
gave different values of light velocity[12]. The light velocity reckoned in the experiment consistent with the procedure A was the mean of two different velocities "to" and

"fro" in respect to the reflecting mirror. From this point of view Einstein did not provide SR with any correct measurement procedure: the procedure A was a classical procedure, defined by adopting the light postulate and it abstracted the changes of length and frequency to which the instruments in motion were subjected. But the procedure B taking these changes into consideration led to the results which were different from the ones of the procedure A. Ives pointed out that Einstein had identified "velocity" and "quotient" which in fact were the operationally different quantities.

In his works[13] Ives considered that measurement of light velocity by the A and B methods and showed the difference between the Newtonian classical notion of velocity and operational notion of quotient. According to one of the Ives' kinematics /IK/ postulates the light is a perturbation of transmitting medium – ether, which is propagated in it with a finite velocity c, irrespectively of the notion of light source: velocity c is defined as the ratio of the distance by stationary sticks /stationary in respect to ether/ to the time interval which is necessary to cover this distance, and which is determined by a stationary clock.

According to the next postulate of IK real sticks and clocks used for velocity measurement are distorted if compared to stationary instruments in proportions:

$$l = l_0 \sqrt{1 - (v/c)^2} \text{ for sticks, } t = t_0 / \sqrt{1 - (v/c)^2} \text{ for clocks,}$$

where l_0 and t_0 are the quantities measured by stationary instruments and V is the velocity determined by instruments in motion relatively to ether.

Ives considered[14] the measurement of light velocity by real instruments assuming that c was a value of this velocity, which is valid only in the reference frame of

ether. It is clear that for the light velocity determined
by the method A the formula defining the velocity has the
same form as the one in regard to ether. In the procedure
A the velocity c and the quotient Q are identical

$$c = 2D/t = 2D'/t' = Q_A ,$$

where D /and D'/ is the length of the distance from the
source of light to the reflecting mirror and t /and t'/ is
the time interval covering the way of light "to" and "fro".
D, t are measured by stationary instruments and D', t' -
by real ones i.e. being subjected by a distortion.

The situation is different in the method B. It appears
that the quotient Q_B determined for light does not equal c:

$$Q_B = c/ \left[1 - c/q \left(\sqrt{1 + (q/c)^2} - 1 \right) \right] ,$$

where $q=D'/\tau'$, D' denotes the length of light trajectory
and τ' the time interval, measured by moving stick and
clock. Q_B does not depend on the velocity of the reference
frame in which the measurement has been performed. So, the
principle of relativity is preserved. Limiting ourselves
to infinitesimally slow motion of the second clock one can
obtain an analogy of the situation recognized in the procedure A: only in the case of such a limited problem of
synchronization one gets $Q_B \rightarrow$ c. According to Ives the
light postulate of the Einsteinian SR consists in identification of Q_B with Q_A, which is to be unacceptable from the
point of view of the discussion on the rigoristic opera-
tionism/operational procedures in physics/.

In the light of the Ives' kinematics SR was operationally incomplete and according to Ives the distinguishable
role of the universal velocity c was not justified in the
sense of finding the appropriate operational procedures.
The constant c played the role of a certain standard of
measurement procedure. Thus its function resembled the role
of the absolute reference frame in the Newtonian mechanics

and the Ives' kinematics.

3. OPERATIONISM IN THE NEWTONIAN PHYSICS AND SPACE – TIME DIMENSION

The Galilean space-time is the space-time for the Newtonian physics. Because the light postulate is characteristic only for the relativistic physics there is not any simple procedure of clock synchronization in the Newtonian physics: there is not any material standard of velocity. However in this case it is also possible[15] to introduce the operationistic procedures of basic kinematic concepts, i.e. time- and distance interval.

According to the procedure A of the clock synchronization in SR the distance d from the point of signal emission to the point of its reflection is determined as follows

$$d = c \left(t_2 - t_1 \right) / 2 \ ,$$

where t_1 is the time of signal emission and t_2 is the time of signal return to this point. The clock synchronization consistent with the light postulate identifies simply two time intervals

$$t - t_1 = t_2 - t \ .$$

This condition in the space of clasical physics should be replaced by a general one

$$t - t_1 = a \left(t_2 - t \right) \ , \ a \text{ is a constant.}$$

Similarly the distance interval can be generalized

$$d = b \left(t_2 - t_1 \right) \qquad . \ b \text{ is a constant.}$$

Two last formulae allow to define the Galilean coordinates in the operational way[16]:

$$t = \left(c_1 t_1 + c_2 t_2 \right) / \left(c_1 + c_2 \right), \ d = c_1 c_2 \left(t_2 - t_1 \right) / \left(c_1 + c_2 \right),$$

where $c_1 = b(1+a)/a$, $c_2 = b(1+a)$.

Contrary to the case of SR the constants c_1 and c_2 are dependent on the observer: if one observer is in motion with velocity v in respect to a second observer who is measuring c_1 and c_2 the values c_1' and c_2' measured by the first observer are the following: $c_1' = c_1 - v$, $c_2' = c_2 + v$. It is easy to check that in the presented operational formalism of classical kinematics $c_1' + c_2' = c_1 + c_2$ in all the inertial reference frames. This invariant has its counterpart in SR: it is the light velocity c. The SR invariant c is given in the form of an axiom while in the classical kinematics each observer should be aware of one more parameter to be able to determine two unknown constants c_1 and c_2. The search for this additional parameter leads to the one-parameter extension of the Galilean group[17] of space-time automorphisms. This control parameter

$$\Theta = \left(c_1^{\,2}(\alpha_2) + c_2^{\,2}(\alpha)_1 \right) / 2,$$

where $\quad \alpha_1 = c_1 t_1 / (c_1 + c_2)$, $\alpha_2 = c_2 t_2 / (c_1 + c_2)$,

is in this case an additional /fifth/ space-time coordinate which informs about the kind of signals needed for operational definitions of d and t. Thus if the operationism requirement is to be satisfied in the classical kinematics it is necessary to extent the space-time to a five dimensional manifold.

4. ON A PARADOXICAL COMPLEMENTARITY OF THE OPERATIONISM REQUIREMENT AND ABSOLUTE ELEMENTS IN KINEMATICS

The contemporary scientific revolution is usually considered as a consequence of the positivistic turn in philosophy. It is claimed that the positivistic requirement to remove from physics all the metaphysics and all the absolute elements which were relicts of the Newtonian physics, essentially changed the structure of new theories as well as the empirical content of their basic concepts. The absolute Newtonian space-time or its XIXth century realizations in the form of ether, became an anachronism of the dispensable metaphysics.

It is no doubt that the positivistic ideas threw a new light on the scientific concepts, the consistency of theoretical structure, and the necessity of mathematical formulation of a theory. In this sense the positivism actually influenced the XXth century science. However it is very important to realize which of its postulates has not been filfilled, to recognize reasons of this failure, and to appraise a possibility of realization of positivistic ideas in physics.

We presented the examples of attempts of operationism postulate realization in kinematics. The old Newtonian dream to derive the theory only from facts, now formulated as the requirement of only empirical concepts, failed once more. The non-observable elements such as the absolute space turned out to be inseparable from any theory in question, which pretended to be just the operationistic one.

The presented analysis of the operationistic completness of kinematics leads to the following conclusions:
1. The relativistic kinematics in the Minkowski space
   formulated as SR contradicts the operationism postulate. The weaker form of the operationistic condition,which bases on the measurement procedure A and the light postulate is realized in SR. From the point of view of IK the light postulate of SR replaces the absolute reference frame for motions and measurements of IK.
2. If the kinematics in the Minkowski space is to be fully
   operationistic it should be of the IK form which conta-

ins the absolute reference frame /ether/ in respect to
which the velocity of light is equal c and the measurement instruments are not subject to any distortion.

3. The clock synchronization and the operationistic definitions of kinematic concepts are not peculiar only
for the space-time with Poincaré symmetry group. It is
also possible to define operationally the concepts of
classical physics /with the Galilean symmetry group/ by
means of the extension of space-time to a five dimensional manifold. In the light of this extension the
space-time manifold of SR is its special simple case.
The additional extension cannot be interpreted in an
operationistic manner.

REFERENCES

[1] Bridgman, P.W., "The Logic of Modern Physics", 1927;
"The Nature of Physical Theory", 1936;"Operational Analysis", Philosophy of Science V, 1938 ; "Some Implications of Recent Points of View in Physics", Reveue
Internationale de Philosophie III, 1949; "The Operational Aspects of Meaning", Synthese VIII, 1950-1;"The Nature of Some of Our Physical Concepts", British Journal
for the Philosophy of Science I-II, 1951.

[2] The operationism in psychology is disscused in e.g.
Israel, H., and Goldstein,B., "Operationism in Psycholo-
gy", Psychological Review LI, 1944.
The operationism in sociology is discussed in e.g.
Lundberg,G., "Operational Definitions in the Social
Sciences", American Journal of Sociology XLVII, 1942;
Dodd,S., "Operational Definitions Operationally Defined",
American Journal of Sociology XLVIII, 1943.

[3] Zahar,E., "Mach, Einstein, and the Rise of Modern Scien-

ce", British Journal for the philosophy of Science 28, 195-213, 1977 .

[4] cf Heller,G., "Ernst Mach", 1960 .

[5] cf Bridgman, 1936; op. cit. in 1.

[6] Zahar, 1977, op. cit. in 3, p. 199.

[7] Mach,E., "The Principles of Physical Optics", 1913.

[8] The absolute element of theory is such an element which cannot be defined in an operational way.

[9] Grabińska,T., "Analiza operacjonistycznych założeń kinematyki H.E. Ivesa i kinematyki relatywistycznej", Z Zagadnień Filozofii Przyrodoznawstwa i Filozofii Przyrody VIII, 7-34;/in Polish/, 1986 .

[10] Kapuścik,E., "The Newtonian Form of Wave Mechanics", Raport Nr.1260/PL, Institute of Nuclear Physics, Cracow, 1984; "On the Physical Meaning of the Galilean Space-Time Coordinates", Raport Nr 1295/PL, Nuclear Physics Institute, Cracow, 1985.

[11] cf Einstein,A., "Zur Elektrodynamik bewegter Koerper", Annalen der Physik 17, 891-921, 1905; "The Meaning of Relativity", Princeton University Press, Princeton,1946.

[12] Ives,H.E., "The Measurement of the Velocity of Light by Signals Sent in the One Direction", Journal of the Optical Society of America /JOSA/ 38, 879-84, 1948.

[13] Ives,H.E., "Genesis of the Querry Is there an Ether?", JOSA 43, 217-218, 1953; 1948, op. cit. in 12.

[14] Ives, 1948, op. cit. in 12.

[15] Kapuścik, 1985, op. cit. in 10.

[16] Kapuścik, 1985, op. cit. in 10.

[17] Kapuścik, 1985, op. cit. in 10.

Author Index

Abers, E.S., 67
Adanyi, A., 8, 9, 12
Aepinus, F.T., 118
Alexander, S., 101
Alhazen, 2, 3, 4
Ampére, A.M., 146
Amsterdamski, S., 66, 68
Angström, A.J., 87, 94
Apianus, P., 4
Apponyi, J., 8, 10, 13
Archimedes, 2
Aristotle, 2, 15-17, 26,
    96, 101-102, 134, 150,
    153
Arp, 183

Bachelard, G., 133
Bacon, F., 104, 115, 131
Bacon, R., 115
Bates of Malines, H., 2
Beaumker, C., 6
Beltrami, E., 76
Bentley, R., 20
Bergson, H., 101
Berkeley, J., 103, 190
Bernoulli, D., 147
Bessarion, J., 2
Białobrzeski, Cz., 172
Bianchi, L., 80

Bieganowski, L., 7
Bielski, A., 7
Birkenmajer, A., 6
Black, J., 119
Blasius of Palma, 4
Beothius, 17
Bohr, N., 114, 124, 140, 145
Boltzmann, L., 146
Bolyai, J., 76
Bondi, H., 183
Bonnet, O., 80
Born, H., 114
Boscović, R.J., 11
Boyer, C.B., 82
Boyle. R., 102, 118
Brady, J.L., 69, 90, 94
Brahe, T., de, 5
Braun, W., von, 47
Bridgman, P.W., 193, 194, 203,
    204
Broncel, Z., 193
Broscius (Brożek), J., 6
Brueck, H., 192
Bruno, G., 150
Buchdahl, G., 110, 112
Burchardt, J., 2, 4, 6, 7
Burt, E.A., 111
Butts, R.E., 111, 112

Subject Index

action et distance, 121
Ad Vitellionem paralipomena, 5
A History of the Theories of Aether ..., 70
Almagestum novarum, 6
antropocentrism, 180
anti-cumulativism, 144
anti-dogmatism, 149, 151
anti-mechanicism, 147
antropic principle, 179, 181, 183, 186, 188-189
Aristoteles et Euclides defensus ..., 6
artifical satellite, 47-50
Assertiones ex universa Philosophia ..., 13, 14
Astronomia Nova, 153
astronomy, 116; Newtonian, 8, 9, 11
astrophysics, 183
atomic events, 161
atomic phenomenon, 163
Austrian National Library, 9

Babylonian Tablets, 94
Basel, 4
Benedictines, 9
benzene, 141
Berlin, 4
Bern, 4
Bibliothéque National, 4
Big Bang, 77, 182, 185
biology, 97

non-observable element, 193
non-statement view, 126, 128
Nuremberg, 4

object, 158, 159, 166-168, 179, 189
observation, 3, 19, 90, 159, 161, 167, 173
observer, 160, 166, 168-169, 178, 183, 189, 201
obstacle, 140, 165; cultural, 136; epistemological, 131, 133
occultism, Peripathical, 9
operationism, 194-195, 197, 199, 201-203
Opticae Thesaurus, 4
Optics, 1, 122
optics, 1, 116, 145
orbit, 50; elliptical, 153
Oxford

Padua, 1, 6
paradigm, Aristotelian, 150; Newtonian, 155-156
parascientific studies, 113, 120, 121, 125
Paris, 9
Pembrook College, 4
Peripathical occultism, 9
Perspectiva, 1-3, 5-6
Philosophia Naturalis, 12
Philosophiae Naturalis Principia Mathematica, 19, 22, 26,
    30, 36, 45, 47-48, 60, 72, 74, 83-84, 107, 111, 117,
    122, 131
philosophy, Aristotelian-Thomistic, 149, 152; experimental,
    101; Goethe´s, 96; Hegel´s, 96; of Nature, 96-102, 110;
    of science, 96-99, 102, 106, 108, 162; Platonic, 3, 72;
    Poincaré's, 126, 127; Schelling´s, 96
phosphorus, 127
Phisica (Physics), 11, 16
Physica Generalis ..., 14
physical world, 123

www.ingramcontent.com/pod-product-compliance
Lightning Source LLC
Chambersburg PA
CBHW050639190326
41458CB00008B/2337